Animal Feeding and Production in India

Animal Feeding and Production in India

Editors
Vir Singh
Ashoka Kumar
R.S. Jaiswal

2011
DAYA PUBLISHING HOUSE®
Delhi - 110 002

ISBN 978-81-7035-740-7

Published by : **Daya Publishing House®**
4760-61/23, Ansari Road, Darya Ganj,
New Delhi - 110 002
Phone: 23245578, 23244987
Fax: (011) 23260116
e-mail : dayabooks@vsnl.com
website : www.dayabooks.com

Laser Typesetting : **Classic Computer Services**
Delhi - 110 035

Printed at : **Chawla Offset Printers**
Delhi - 110 052

PRINTED IN INDIA

Preface

Agriculture encompasses all land-related activities, including cropping, animal husbandry, horticulture, poultry, dairying, sericulture, forestry, etc. Domestic animals are the main axis of Indian agriculture. In many areas, livestock are more than mere economic resources for the farming communities. They are the very basis of the culture they are the part of. In other words, we can say that domestic animals are a life-line of Indian culture.

Animal production has bearing on Indian agriculture and animal feeding has bearing on animal production. Animal production and animal feeding are the interlinked processes. Animal feeding, which inevitably depends on available land resources, feed resources, agro-climatic zone, cultivation practices, farmers' socioeconomic status, and the available species and breeds of animals, is the most important aspect of animal production. Animal production, in essence, revolves round animal feeding. Animal production, as critically influenced by animal feeding, spells well-being of a farming community and has enormous impact on national food security as well as on gross national production.

Animal Feeding and Production in India structured jointly by the authors collates some selected articles/ research papers which would have phenomenal impact on livestock-based economies of India. The book especially serves to fulfill the need of livestock-based livelihoods in rural, mountain and marginal areas.

The understanding of principles of animal production and feedings should not be inhibited by the principles of resource management, such as forest/ rangeland resource management. In fact, both natural resource management and animal production are complementary to each other. One is benefitted by the other. Natural resource management, particularly focused on forest/ rangeland resources, helps augment fodder base for the livestock, as well as takes into account the principles of resource conservation. Livestock, after all, are part of agro-ecosystems. Health of an agro-ecosystem and livestock resources goes hand in hand. When we take natural resource conservation principles into account, we essentially apply the principles of sustainability in livestock production. This book attempts to throw some light on sustainable approaches to livestock production and feeding.

Attempts have been made to cover all aspects of livestock production and feeding. All important livestock species – cattle, buffaloes, Angora rabbits, goat, sheep, poultry – which are reared in India and serve rural economies to a great extent have been taken into consideration. Issues such as total mixed rations, nutritive value of cultivated fodders, ethno-medicines, draught animal power, Angora rabbitry, fostering nutritive value of poorly digestible straws, livestock-mediated agro-ecosystem services, rangeland resources, unconventional feeds such as mustard straw, etc. have been intensively discussed. There are frequently asked questions which need to be responded properly. The last chapter of the book beautifully answers such questions.

We are grateful to all the authors and learned researchers for their contributions. We are also grateful to Daya Publishing House for publishing the book.

When this book was being structured, a bad news poured in—demise of Dr. Mahendra Singh Rahal, former professor of Animal Nutrition at the GB Pant University of Agriculture and Technology, who has been an icon and a celebrity amongst animal nutritionists. Dr. Rahal searched many new horizons in the field of animal nutrition. His contributions to the field of animal nutrition are unparallel and exemplary. The editors of the book have been proud students of Dr. Rahal. We deeply mourn Prof. M.S. Rahal's death and dedicate this book to his memory.

While every care has been taken during the editing of this volume, we would also like to invite readers'/ researchers' critical comments and suggestions for further improvement.

Vir Singh

Ashoka Kumar

R.S. Jaiswal

Contents

List of Contributors

Ashoka Kumar, Department of Animal Nutrition, College of Veterinary and Animal Sciences, GB Pant University of Agriculture and Technology, Pantnagar-263145, Uttarakhand.

Babita Bohra, International Centre for Research in Agroforestry (ICRAF), National Agricultural Science Center, New Delhi.

C.B. Singh, Department of Genetics and Animal Breeding, College of Veterinary and Animal Sciences, GB Pant University of Agriculture and Technology, Pantnagar-263145, Uttarakhand.

Chetna Bhatt, Department of Animal Nutrition, College of Veterinary and Animal Sciences, GB Pant University of Agriculture and Technology, Pantnagar-263145, Uttarakhand.

D. Kumar, Department of Genetics and Animal Breeding, College of Veterinary and Animal Sciences, GB Pant University of Agriculture and Technology, Pantnagar-263145, Uttarakhand.

D.N. Pandey, Department of Animal Nutrition, College of Veterinary and Animal Sciences, GB Pant University of Agriculture and Technology, Pantnagar-263145, Uttarakhand.

Kalpana Joshi, Department of Animal Nutrition, College of Veterinary and Animal Sciences, GB Pant University of Agriculture and Technology, Pantnagar-263145, Uttarakhand.

Kusum Rikhari, Department of Animal Nutrition, College of Veterinary and Animal Sciences, GB Pant University of Agriculture and Technology, Pantnagar-263145, Uttarakhand.

Mahendra Singh, Department of Animal Nutrition, College of Veterinary and Animal Sciences, GB Pant University of Agriculture and Technology, Pantnagar-263145, Uttarakhand.

Manisha Joshi, University of Manitoba, Winnipeg, Canada.

R.D. Gaur, HNB Garhwal University, Srinagar, Pauri Garhwal, Uttarakhand.

R.S. Jaiswal, Department of Animal Nutrition, College of Veterinary and Animal Sciences, GB Pant University of Agriculture and Technology, Pantnagar-263145, Uttarakhand.

Richa Pathak, Department of Animal Nutrition, College of Veterinary and Animal Sciences, GB Pant University of Agriculture and Technology, Pantnagar-263145, Uttarakhand.

Ripusudan Kumar, Research Station, GB Pant University of Agriculture and Technology, Majhera, Nainital, Uttarakhand.

Tej Partap, Vice-Chancellor, Sher-e-Kashmir University of Agricultural Sciences and Technology (SKUAST), Srinagar, J&K.

Vir Singh, Department of Environmental Science, CBSH, GB Pant University of Agriculture and Technology, Pantnagar-263145, Uttarakhand.

Woda Jeremaih Odok, Woda Jeremaih Odok, Assistant Professor, Upper Nile University, Khartoum- P.O. Box 1660, Sudan

Chapter 1

Feeding of Animals

☆ *Ashoka Kumar*

Dairy farmers generally aim at maximizing milk production in a cost-effective manner. On a typical dairy farm, feed costs represent nearly 50-70 per cent of the total milk production costs. Maximum feed intake, improved conversion efficiency and lower feed costs are the key factors for high economic returns from a dairy farm. With high producing cows, successfully implementing management strategies to maximize feed intake will determine how a balanced diet supports milk production.

A dairy farmer must take it for granted that one of the most important parts of dairying is scientific and sound feeding of animal. The feed and fodder should be nutritive, fibrous, scientifically balanced, economical and palatable.

Nutrient Requirements

Heredity limits the very ability of a dairy animal to produce milk. Difference in level of milk production among cattle is determined by approximately 25 per cent heredity and 75 per cent environment. A large part of the environmental factor that influences milk production is the feed available to dairy cattle.

Basis of Scientific Feeding

1. Knowledge of the quantitative needs of the animal body for the nutrients.
2. The relative value of feeds as a source of these nutrients.

All the living organisms require specific essential nutrients to satisfy the biological processes associated with:

1. Growth of immature animal.
2. Pregnancy needs are small during the first six months and large during the last two or three months of the gestation period.
3. Fattening or regaining of normal body weight lost during lactation.
4. Maintenance of the mature animal: Needs vary according to size of the animal.
5. Milk production: Needs varying with the kg of milk produced and its composition.

The use of adequate and well balanced diets can:

1. Maximize profits.
2. Minimize losses in a scientific feeding programme.

An animal's diet must contain the essential nutrients in appropriate amounts and ratios.

The important issues in scientific livestock feeding are:

1. Feed/ration that is properly balanced scientifically for all the essential nutrients and is palatable (good taste).
2. Maximum feed intake and better feed conversion efficiency.
3. Crude protein makes up 14 to 19 per cent and carbohydrates 65 to 75 per cent of the total ration.
4. DM in common dairy cow's diets to optimize solid-corrected milk production.
5. A ration should be balanced for NDF, NFC, Crude protein, and escape protein (rumen bypass protein).
6. Producers must always be aware that the nutritional requirements vary with the species of animal, age of animal, stage of production and even the season of the year.

Methods of Feeding Dairy Cattle/Animals

A. Traditional

Roughages are fed free choice in bunks to the entire herd or in mangers. Concentrates are fed to cows individually, according to milk production/growth. The concentrate mixture is fed in mangers or in the milking parlor during milking.

Advantages of Traditional Feeding

1. Less specialized equipment is needed.
2. Theoretically feeds each animal/cow according to individual needs based on production.
3. Permits adjusting concentrate feeding to the stage of lactation.
4. Permits challenge feeding each animal.

Disadvantages of Traditional Feeding

It is difficult to measure the actual amount of forage (roughages) each head of the cattle eats. Therefore, it is difficult to balance the ration with the right amount of concentrate for each animal.

1. Low producers are often overfed concentrates.
2. High producer are often underfed concentrates.
3. The level of dust in the milking parlor increases.
4. Milking in parlor may be delayed while waiting for the cow to finish eating the grain mixture.
5. Feeding in the parlor slows down the milking.
6. More labourers are required.
7. There is less control over the total feeding programme.
8. Cost of equipment is high.

Careful records of individual production and continual adjustment of concentrate feeding to match production are required.

B. Challenge or Lead Feeding

It is the practice of feeding higher levels of concentrate to challenge the cow to reach the maximum of her potential milk production. Challenge feeding gives more concentrates to the cows early in the lactation period and less during the later part of the lactation period. Feed dry cows in good condition about 0.488 kg of concentrate per 100 kg of body weight. Two or three weeks before calving, increase the amount of concentrate by about 0.45 kg (450 gm.) per day until the cow is eating from 1.0 to 1.5 kg per 100 kg of body weight at calving. About three days after calving, increase the concentrate by 1.0 to 2.0 kg per day for two to three weeks after calving. At this time, the cow should be eating about 2.0 kg of concentrate per 100 kg of body weight daily. Continue to increase the level of concentrate feeding until milk production levels stop. At maximum milk production hold the level of concentrate feeding constant.

Increase the level of concentrate feeding during the early part of lactation period if the cow is losing weight. Cows that do not respond to challenge feeding should be culled from the herd.

C. Feeding Total Mixed Ration (TMR)

Total Mixed Ration (TMR) is a ration in which all feedstuffs are blended together to be fed as one feed. A properly managed TMR will ensure that all cows are receiving a balanced diet. TMRs help to prevent many common nutritional and metabolic disorders. TMRs allow feed to be used more efficiently by dairy animals. There are many advantages and few disadvantages to TMRs.

Quality of Concentrate Mixture

The quality of feed ingredients determines the quality of compounded feeds or finished products. So, it is very much essential to have a standard for every feed ingredients including the supplements and feed additives.

For determination of feed quality, the following points are taken into considerations:

1. Physical characteristics of feed ingredients
2. Chemical composition of feed ingredients

Physical Characteristics of the Feed Ingredients

The assessment of physical form of feed is made on the basis of some observations, which are as follows:

1. Soundness of grains, seed, or any feed ingredients
2. Dampness

3. Colour of the feed ingredients
4. Texture of the feed ingredients
5. Presence of cake formation and extent of caking
6. Presence of extraneous materials in the feed
7. Infestation (insect and/or fungal)

Soundness of Grain, Seed and Other Feed Ingredients

A. The feed grain, seed, etc. should be free from cut, touched and heat damaged grain, thin grains and foreign materials.

B. A tentative Indian grading system of cereal grains is vague and more advantageous for these purposes.

Grading of Cereals

Grade No.	*Percentage of Cut, Broken, Touched, Discoloured and Heat Damaged Grains*
1.	< 5.00
2.	5.1 to 10
3.	15.1 to 20

Dampness

This is determined by taking random samples from different parts by hand picking. A grain damped for more than a day gives off flavour.

Colour of the Feed Ingredients

Feed ingredients impart characteristic odour, which is very light in feed grain but sensible in most of the oil cakes and leaf meals. The odour of *rape* and *mustard* is pungent, *ground nut cake* is acceptable and *seasam cake* is slightly off in odour.

Texture of Feed Ingredients

The grains and other feed ingredients should be sound and free from impurities, dampness and caking.

Extent of Caking

The presence of few lumps or cakes in feed ingredients may be due to accidental dampening and may not cause serious quality problems but excessive lumping with or without fungal infestation must be tested for mycotoxin before its acceptance.

Presence of Extraneous Materials

The percentage of foreign materials is determined by screening and then the feed is graded. The presence of toxic weeds and other materials should be examined carefully without any negligence.

Infestation

Insect infestation has been made acceptable to a greater degree but fungal infestation resulting in higher than 25 ppb of aflatoxin are not acceptable.

Chemical Composition of Feed Ingredients

Routine feed analysis should be done to determine its quality. It is analysed for moisture percentage and other proximate composition (protein, ether extract, nitrogen free extract, crude fibre and total ash) and acid insoluble ash content in feed ingradients. Now these values are compared with the standard values and accordingly the quality is determined. For certain feeds some specific analysis is also done like urea in protein supplement, sand in mineral mixture, etc.

The sample of mineral mixtures are analyzed for contents of calcium, phosphorus, copper, cobalt, manganese, iron, sulphur as minimum levels of most of these elements have been fixed in a standard mineral mixture. Similarly, the maximum levels of fluorine and selenium are also fixed in the mineral mixtures.

Determination of carotene and some other vitamins may be required in compounded feeds for different kinds of animals like poultry and pigs.

Evaluation of Animal Feed Quality

The animal feeds mainly consist of two types of feed materials *i.e.* concentrate and roughages. It is essential to monitor the quality of these materials to provide the animal a sound health and production.

1. Evaluation of Quality for Concentrates

The concentrates include cereal grains, beans, different types of oilcakes, animal origin protein supplements and molasses etc. The feed ingredients are carefully inspected at the time of purchase and/or selection for the feeding of animals. Following points are generally considered for the appraisal of the feed ingredients:

(a) Inspection of Dryness

The sound and dry grains possess characteristic bright color. The grains easily get scattered in larger area on throwing on dry surface. It is hard to crack a dry kernel or seed between the molars and on cracking it produces characteristic sound and splits into two or more particles without flaking or flattening.

(b) Detection of Impurity

Many times food grains may contain a related edible grain and its presence may be intentional or unintentional. Higher content of such type of impurity may require consideration, *e.g.*, an impurity of barely or oat up to one per cent may be ignored.

(c) Detection of Weed Seeds

The seeds of weeds may or may not be injurious for health and production of the animals. A few seeds of some weeds may be tolerated but in larger proportion become injurious to health. Weed seeds like lupins, wild peas, and lathyrus may be accepted upto one per cent level in feed ingredients as this amount can be tolerated by the animals but weeds like Dhatura seeds, Castor seed etc. should not be present in animal feed as it is highly toxic to animals.

(d) Detection of Extraneous Particles

It includes grits of stone and gravels which may be present in animal's feed that may not be very harmful for animals but will cause problem in feed grinding.

(e) Detection of Insects and Weeviling

The lot of grains selected for feeding of animals should be free from live or dead insects and their eggs. There should not be any weeviled, touched or germ eaten grain in the bulk. If the percentage of discolored, damaged and touched grains in representative sample is more than 20 per cent, then such lots should be rejected for the use as animal feeds.

(f) Physical Appearance

Physical appearance should be characteristics for particular feed ingredients, like wheat bran is flaked, bulky and pinkish grey. Til cake from whole seed extraction in flack. Mustard cake appears greenish brown or dark green depending on variety where as ground nut cake is grey.

(g) Odour and Texture

Odour should be characteristic, like it is some what sweet for groundnut cake, pungent for mustard cake. Ghani and expeller pressed cake are lustrous due to presence of more oil as compared to solvent extracted cake which is generally dull in texture.

2. Evaluation of Quality for Roughages

Dry roughages commonly procured for feeding of farm animals include wheat straw, rice straw, sorghum kadbi, millet kadbi and maize stover etc.

Wheat Straw

It is mainly available in the form of bhoosa. It is golden yellow in color and bright in appearance. Its dust content should be less than one percent. It should be free from cake formation and fungal infestation.

Paddy/Rice Straw

It is generally available in long form with highly variable size and colour. The colour ranges from pale yellow to golden yellow. It should be free from lumps of clay and fungal infestation.

Sorghum Kadbi

It is light green to light brown in colour and should be free from fungal infestation.

Millet Kadbi

It is brownish grey in colour and it is highly susceptible for ergot infestation which should be checked before procurement.

Maize Stover

The colour of maize stover is brown to greenish yellow. It should also be free from clay and fungal infestation.

Feeding Management of Calf from Weaning to Heifer to Breeding Age

Proper diet is necessary if the heifers are to be ready for breeding purposes at the right time. Improper diet during this period of growth will result in heifers not producing milk to their potential when they are in milking herd.

Colostrum milk is the first milk secreted by the cow/buffalo after calving. Feed the colostrums milk within 30 minutes after birth if possible, and certainly not later than four hours after birth. Colostrum milk, as compared to normal milk, is high in fat, solid-not-fat, total protein, and antibodies, which protect against diseases.

The newborn calf does not have enough antibodies to protect it from diseases until it receives the colostrums milk. The ability of the calf to absorb the disease-protection component (gamma globulin) of colostrums milk is sharply reduced after 24 -36 hours. The danger of infection from disease-causing bacteria is high during the first hours after the calf is born. Early feeding of colostrums milk will reduce death losses of newborn calves. Wash the cow's udder and teats before allowing the calf to nurse. This will reduce the chances of calf picking up germs and dirt while nursing.

Usually calves search udder (teats) for suckling milk after ½ to 1 ½ hours after the birth. They stand and move towards the udder of dam under natural rearing practices followed. Under weaning system, calves start suckling ear, navel, scrotum, dew-lop or any part of the body of other calves, if kept together. Therefore, one should be careful while rearing calves under weaning system. Calves are fed milk from their first day of life by two ways:

1. Nipple feeding
2. Bucket feeding

Under nipple feeding system milk is filled in bottles with nipple. The temperature of the milk is at body temperature, *i.e.* 100 to 101°F. The nipple is put near the mouth of calves. They start suckling by natural instinct. But under bucket system of feeding calves are to be trained for feeding milk from bucket or pail.

Feeding Routines

In first three weeks of life diarrhoea occurs due to over feeding. Therefore, calves should be fed twice or thrice per day at an interval of 8 to 12 hours, *i.e.*, the total amount of milk required is fed into divided doses. The calf should be fed colostrums milk for the three days after birth. During this time, colostrums milk is higher in nutrients than regular whole milk or milk replacers. After three days, the calf may be fed whole milk, excess colostrums or milk replacer. Small breed calves are fed about 2.3 kg and large breed calves are fed about 3.2 kg of whole milk, excess colostrums, or milk replacer. The amounts are fed for three weeks on an early weaning programme and for four or five weeks on liberal milk feeding programme.

Milk replacer is usually fed in the gruel form. The amount is gradually increased with simultaneous decrease in amount of whole milk. It should have good quality ingredients.

Formulation of Milk Replacers

Ingredients	Proportion (%)
A.	
Dried skim milk	50
Dried whey	30
Dextrose	8
Oat flour	5
Brewer's yeast	5.26
Trace minerals	0.04
Vitamin A supplement	1.7
Total	**100**
B.	
Wheat flour	10
Fish meal	12
Linseed meal	40
Coconut oil	7
Linseed oil	3
Butyric acid	0.3
Citric acid	1.4
Molasses	10
Mineral mixture	3
Aurofac	0.3
Milk	13
Total	**100**

Use milk replacers of high quality, made from dairy products rather than from plant products. Milk replacers that have the following protein sources are preferred:

- ☆ Skim milk
- ☆ Buttermilk powder
- ☆ Dried whole whey
- ☆ Delactosed whey
- ☆ Casein
- ☆ Milk albumin

When all protein sources are from dairy products, the milk replacer should contain 20 per cent protein. When plant protein sources are used in the milk replacer, the protein content should be 22 -24 per cent.

A minimum fat level of 10 per cent is recommended in milk replacer and the level may be as high as 30 per cent. Fat reduces the scours and provides needed energy in the diet.

The amount of milk ıequired for the calf is as follows:

Age of Calf (weeks)	*Colostrum (kg)*	*Milk (kg)*	*Whey Milk (kg)*	*Concentrate Mixture (gm)*	*Hay/Green Fodder (kg DM)*
Frist 3 days	2–2.5	–	–	–	–
Remainder days:					
1	–	1.5–3.0	–	50	0.2
2	–	3	–	100	0.35
3	–	3.25	–	200	0.4
4	–	3	–	350	0.5
5	–	1.5	1	600	0.55
6	–	–	2.5	600	0.6
7	–	–	1.75	700	0.75
8	–	–	1.25	800	0.85
9	–	–	–	900	1
10	–	–	–	1000	1
11	–	–	–	1000	1.2
12	–	–	–	1200	1.3
13	–	–	–	1500	1.5

Therefore, as per the requirement, milk can be fed to the calves for keeping them healthy and in proper health with optimum growth rate. It is important that calves get enough dry matter diet for proper growth.

Milk or milk replacer may be fed with bucket or nipple pail. Take care to keep the bucket or pail clean for each feeding. A dairy bucket will increase the chances of the calf getting scoures.

Calves must be healthy and not subjected to other stress conditions such as cold weather if they are to be successfully weaned before four weeks of age. Calves may be weaned at four to five weeks of age if they are eating well and are healthy. During the last week, reduce the amount of milk fed 1.4 kg for small breeds and 1.8 kg for larger breeds.

Milk or milk replacer may be fed up to six or eight weeks of age, but this will increase the cost of rearing the calves. Reduce the amount being fed during the last two or three weeks before weaning.

Calf starter is a solid feed consisting of ground grains, oil cakes, animal protein supplements and brans. Calf starter should be fed in the beginning when the calf is about four days of age. Teach the calf to eat the starter by rubbing a little on its nose after each milk feeding. The calf will soon start to eat the starter. After the calf begins to eat the starter, it may be fed free choice. The calf should be eating 0.45 to 0.9 kg by the time it is four weeks of age. Feed up to 1.4 kg to small breeds and 1.8 to larger breeds per day. Feed calf starter for three or four months.

Composition of Calf Starter

Feed Ingredients	*Proportion (per cent)*
Crushed barley	50
Wheat bran	08
Groundnut cake	30
Fish meal/dried skim milk/meat meal	10
Mineral mixture	02

To 100 kg of above mixture the following ingredients may be added:

- ✰ Molasses 5-10 per cent
- ✰ Rovimix 10 gm/quintal

- ☆ Common Salt 0.5 per cent
- ☆ Aurofac 20 gm/quintal

Calf starters are palatable and must contain about 16 per cent crude protein if the calf is to be weaned early (before four weeks); the calf starter should contain 18 per cent crude protein. Good calf starters contain whole, coarsely ground, cracked grains. Palatability may be improved and dust reduced by adding up to 5 per cent molasses in the mixture.

High quality hay may be fed at the beginning of about four weeks of age; however, forages are not necessary in the ration until 8 to 10 weeks of age. If forage is not fed early, it is recommended that additional fibre be included in the starter ration. Hay is a better roughage for young calves. Do not feed moldy or damaged hay to calves. Bright, leafy, early cut legume-grass hay makes good roughage for young calves. Feed the hay free choice.

Feeding the Calf from Weaning to One Year of Age

The ration can be forage fed free choice with 1.8-2.3 kg of grain/concentrate mixture per day. The amount of protein supplement needed depends on the protein content and amount of forage fed. Fodder may supply some of the roughage needed but will not supply all the nutrients needed for growing heifers. It is recommended that heifers on fodder be fed some grain/concentrate mixture and stored forage to supplement the nutrients provided by the fodder. Trace-mineralized salt and calcium phosphate mineral mixture can be fed as licks if the grain mixture does not meet these needs. A supply of clean, fresh water is also essential.

Care must be taken not to allow heifers to become too fat. It may be necessary to limit the amount of grain in diet to prevent this.

A 12-16 per cent protein level in the forage will eliminate the need for a protein supplement fed with the grain/concentrate mixture. The same grain/concentrate mixture that is fed to milking herd may be used if it contains enough minerals and vitamins.

Feeding Heifers One to Two Years of Age

A good quality fodder may supply all the nutrients needed during this period of growth. Feed trace-mineralized salt and calcium-phosphorus mineral mixture as licks. A concentrate mixture

1.0–1.5 kg per day is desirable. It may be necessary to feed some additional concentrate mixture to maintain this rate of gain.

If fodders are mature it will be necessary to provide additional feed. A diet deficient in energy, phosphorus, or vitamin A will prevent the onset of estrus.

There is a relationship between the amount of energy in the ration of developing heifers and reproduction. A shortage of energy in the ration delays the time the heifer reaches first heat, which causes delays in breeding and shortens the productive life of cattle/ buffaloes.

High-energy intake during growth will cause the heifer to reach first heat earlier, but may result in breeding problems when the cow/ buffalo is mature. Cows overfed energy have a higher sterility rate than cows fed the right amounts of energy/basal feeds. Overfeeding energy also shortens the productive life of the cow, and bulls that become too fat have problems with sperm production.

A shortage of phosphorus in the ration appears to cause irregular heat cycles; therefore, it is important to meet the cow's needs for phosphorus. Breeding problems can also be caused by a shortage of vitamin A in the diet. The addition of vitamin A to the ration will help prevent these problems.

General Feeding of Lactating Dairy Cows

The importance of balancing rations for dairy cattle to maintain a high level of production is indicated by the fact that the feed costs represent approximately 50-60 per cent of the total cost of producing milk. The ability of a dairy cow to produce milk is limited by heredity. Differences in the level of milk production among cows are determined by approximately 25 per cent heredity and 75 per cent environment. A large part of the environmental factor that influences milk production is the feed available to the cow.

The diet of the dairy cows has a major influence on their ability to produce milk according to genetic potential. Maintenance and milk production needs are the two most important to consider when balancing rations for dairy cows. With high producing cows, nutrient requirements for milk production may be several times that needed for maintenance. Rations for lactating dairy cattle are generally balanced for energy and crude protein.

Dairy cattle are ruminants and can use large amounts of roughages in the ration. Use high quality roughages and supplement the ration with concentrates (grain and protein feeds). Minerals needs of dairy cattle are met by some mineral supplement to the ration and also by providing a mineral supplement as licks.

Roughages are fed free of choice in bunks/manger for the entire herd. Concentrates are fed to cows individually, according to the milk production. The concentrate is mixture fed in milking parlor during milking. The most common method of feeding dairy cows today is by using a complete ration. The complete ration has all or almost all of the ingredients blended together and is fed free of choice to all the cows of a group. The roughages and concentrates are required to meet the energy, protein, mineral, vitamin, and crude fibre needs of the cows are blended together in a complete ration. When feeding a complete ration free of choice, no concentrates are fed in milking parlor.

Feeding of Young Dairy Stock (Calf Feeding)

The first step in production of dairy calves is to have strong calves at birth. To ensure this, the cows should have rest in last two months of dry period between successive lactation and should receive a well balanced ration for regaining their body weight, if there is any loss in last lactation.

Under nourished, stunted calves often, although not always may be grown out into large animals by carefully feeding. The higher the plane of nutrition, the earlier is the onset of puberty and thus the quicker the return of capital.

Guidelines for Feeding Lactating Dairy Cows

A shortage of energy is usually the most limiting factor in milk production. The total ration for lactating dairy cows should contain 60-70 per cent total digestible nutrients (TDN).

Forage is the basic need for a dairy-feeding programme. On an average feed 2.0 to 3.0 kg of forage dry matter per 100 kg of body weight. A minimum level of fibre in the ration is necessary to maintain milk fat percentage in the milk. The ration should contain a minimum of 15 per cent crude fibre. When roughage is fed free choice, daily intake (dry matter basis) as per cent of body weight will vary from 1 per cent for low quality to 3 per cent for high quality forage.

The total ration should contain from 18 -22 per cent CP on an as fed basis. Approximately 75–80 per cent of crude protein (CP) in the ration is digestible and available for use by dairy cow.

Use 0.5 to 1.0 per cent common salt and 1.0 to 2.0 per cent mineral mixtures in the concentrate mixture. These levels will provide the major and trace minerals needed by the dairy cows. These levels also maintain the overall health and improve reproductive efficiency of the cow.

Grains and forages are often processed before feeding; use a coarse to medium grind for dairy cows. The most expensive part of the ration is the grain and protein supplement. The use of home grown grains whenever possible will generally help to lower the cost of the ration. Buy feeds on the basis of the least cost per unit of nutrient provided.

Feed requirements vary with the stage of lactation and gestation. The most critical feeding period is the first ten weeks of lactation, when milk production is increasing rapidly. During the first ten weeks of lactation, increase the amount of grain in the ration by 0.45 -0.90 kg per day. Maintain the fibre level in the ration above 15 per cent to keep the rumen working properly. Limit the grain and their byproducts to more than 65 per cent of the total dry matter in the ration. Do not increase the rate of grain feeding too fast or feed more than the cow should eat. Increasing the rate of grain feeding too fast may result in digestive problems such as acidiosis, displaced abomasums, or going off feed. On the other hand, if the nutrient needs of the cow are not met, problems with low-peak production and ketosis may develop.

A younger cow (2- and 3-year olds) needs extra amounts of feed nutrients for continued growth during lactation. If their nutrient needs are not met, they often will not reach their full milk producing potential.

Milk production peaks approximately six to eight weeks (42 - 48 days) after calving. During the second ten weeks of lactation, feed a ration to maintain peak milk production. Grain intake will equal about 2.5 per cent of the body weight and roughage intake will be about 1 per cent of the body weight. It is important to maintain an adequate level of roughage in the ration to maintain rumen function and milk fat test.

During the latter part of lactation (140–305 days after calving) milk production is usually dropping. Generally, the grain intake should be matched to the level of milk production. However, thin cows will need extra grain in the ration to restore the body condition they had prior to the dry period. It is easier to improve condition during lactation than during the dry period. Younger cows are fed more nutrients to meet their growth needs. Two-year olds should get 20 per cent more and three-year olds 10 per cent more nutrients.

Feeding and Reproduction

A protein shortage in the ration may cause silent heat or discontinued heat. Protein shortage is more common when corn silage is the main forage fed in the ration. Greater care must be taken to properly balance the ration for protein when corn silage is the main forage fed.

A shortage of phosphorus in the ration appears to cause irregular heat cycles; therefore, it is important to meet the cow's needs of phosphorus.

Breeding problems can also be caused by a shortage of vitamin A in the diet. The addition of vitamin A to the ration will help prevent these problems.

Guidelines for Feeding of Dairy Cows

Dry cows normally need fewer amounts of nutrients than lactating cows. Nutrients are needed for developing calf and replace body weight lost during lactation. Care must be taken not to over-feed dry cows or they will become too fat. Limit the intake of corn silage and grain during dry period because of their high energy content. This will reduce problems with excess fat deposit in liver area and displaced abomasum.

Total dry matter intake during the dry period should be limited to two per cent of the body weight. Roughage intake must be at least one per cent of the body weight. The amount of grain needed depends upon the quality and type of the forage in the ration. Grain intake should be limited to not more than 0.50 per cent of body weight daily.

If no grain is fed during the dry period, begin some grain feeding during the last 02 weeks before calving. This will prepare the rumen for digesting grain during the lactation period. Begin grain feeding with about 1.8 kg daily and slowly increase the daily amount fed to a maximum of 5.44–7.25 kg per head at calving time.

Limit calcium intake to 100 gm daily. A minimum 40 gm of phosphorus should be fed daily. The calcium-phosphorus ratio should be about 2:1. Reducing the level of calcium in the diet to less than 0.2 per cent about two weeks before calving will reduce the chances of milk fever developing at calving time.

When poor quality feeds are included in the ration, add vitamin A and D. The addition of these vitamins increases the calf survival rate and reduces the problems with retained placenta and milk fever at calving time.

Trace minerals, especially iodine and cobalt, are included in the ration. If urea is to be fed during lactation, begin feeding it about two weeks before calving, so the rumen has time to adjust to the urea in the ration.

Principles of Feeding Dairy Cattle

Feed represents the major cost in animal production, which may range from 60-80 per cent. Thus, it is imperative to supply an adequate diet in terms of nutrient content and to prepare in a manner that will encourage consumption without waste and allow high efficiency of feed utilization. Feed formulation is an art, which has to ensure least cost and optimum quality.

Feeding of farm animals necessitates the provision of different nutrients in required amount consumable through dy matter by the animal in order to get a particular level of production. No feed is complete and either provides low or high amount of nutrient(s) than the requirement if fed alone. They differ in their physical characteristics as well as in their nutrient supplying capacity to the animal. This necessitates the feeding of more than one type of feed. In efforts to make an adequate balance diet, a number of different feed ingredients are to be mixed together so that desired concentration of one or more nutrients could be achieved in possible dry matter to be consumed by the animals. After knowing the principles of feeding animals, one can easily proceed to formulate the balanced ration for the livestock. The ration so computed should be economically balanced and cheap as far as possible. To formulate a diet/ration, therefore, the following information in respect of nutrients and dry matter intake (DMI) is required:

1. Need of the animal for different functions
2. The availability of different feed ingredients along with their nutritive value.
3. Cost of the feed ingredients is available at hand.

Following are some examples of ration formulation:

Example 1

Compute the balanced ration for a growing heifer weighing 200 kg. The feeds available with the farmer are berseem hay, crushed (dalia) oat and crushed maize.

Solution

First find out the requirement of the heifer in terms of dry matter, DCP and TDN. The DM can be calculated as follows:

For 100 kg body weight the animal can consume 2.0 to 2.5 kg DM.

For 200 kg body weight the heifer can consume 4 to 5 kg dry matter.

So the total requirement of the heifer can be as under:

DM (kg)	*DCP (kg)*	*TDN (kg)*
4 to 5	0.40	3.0

Hence the feeds available should be mixed and supplied in such a manner that the requirement is met properly. The percentage of DCP, TDN and dry matter in the feeds can be seen from the table of the nutritive value of the feeds.

According to the principles of feeding as 2/3rd requirement of the dry matter 2.5 to 3.0 kg should be met from roughages (berseem hay) and remaining 1/3rd (1.5 to 2.0 kg) from concentrates (Oat + Maize). Since the percentage of dry matter in all these feeds is about 90 per cent , hence roughly about 3.5 to 4.0 kg berseem hay and 1.6 to 2.3 kg concentrate mixture made by oat and heifer will need maize. Keeping in view the above facts proceed for calculating the actual amount of the ration of the heifer.

Method 1: Without Formulating Concentrate Mixture

Name of the Feed	*Quantity (kg)*	*DM (kg)*	*DCP (kg)*	*TDN (kg)*
Berseem hay	3.75	3.375	0.337	2.212
Ground oat	0.50	0.450	0.035	0.355
Ground maize	0.75	0.675	0.052	0.637
Total	5.0	4.490	0.424	3.204

Common salt and mineral mixture 25-30 gm per day each should be provided.

Method 2: By Formulating Concentrate Mixture

Name of the Feed	*DM (per cent)*	*DCP (per cent)*	*TDN (per cent)*
Oat 100kg	90	7	71
Maize 100 kg	90	7	85
Total	180	14	156

Therefore 1 kg of concentrate mixture will contain 0.9 kg DM, 0.07 kg DCP and 0.78 kg TDN.

Now the calculation can be done directly as under:

Name of the Feed	*Quantity (kg)*	*DM (kg)*	*DCP (kg)*	*TDN (kg)*
Berseem hay	3.75	3.375	0.337	2.212
Conc. Mixture	1.27	1.143	0.089	0.994
Total	5.02	4.518	0.426	3.206

Common salt and mineral mixture should be given 25-30 gm/d each.

The ration thus computed will be economically balanced, cheap and more suited to the requirements of the heifer. When many numbers of the feeds are given, it is the choice of the computer to select any few of them.

Example 2

Compute the balanced ration for a dry cow weighing 250kg. The feeds available with the farmer are green cowpea and wheat straw.

First Calculate the Requirement of the Dry Cow

Particulars	*DM (kg)*	*DCP (kg)*	*TDN (kg)*
Requirement for maintenance	5.0-6.25	0.168	2.02
% in Cow pea	30	2.2	11
% in wheat straw	90	0.0	40

Name of the Feed	*Quantity (kg)*	*DM (kg)*	*DCP (kg)*	*TDN (kg)*
Green cow pea	7.7	2.31	0.169	0.847
Wheat straw	3.0	2.70	0.000	1.200
Total	10.7	5.01	0.169	2.047

Common salt and mineral mixture should be given 25-30 gm/d each.

Since the leguminous green is available in the ration, there is no need of feeding concentrate mixture to dry cow. Wheat straw is mixed with green cow pea to adjust the appetite.

Example 3

Compute a balanced ration for a lactating cow weighing 400 kg and yielding 10 lt. milk/d with 4 per cent fat. The available feeds with the farmer are wheat straw, green berseem, green maize, wheat bran, crushed gram, barley and groundnut cake.

Requirement	*DM (kg)*	*DCP (kg)*	*TDN (kg)*
Maintenance requirement	8-10	0.254	3.03
Production requirement	2.0	0.450	3.16
Total requirement	10-12	0.704	6.19

Feed Ingredients	*DM (kg)*	*DCP (kg)*	*TDN (kg)*
% in wheat straw	90	0.0	43
% in green berseem	30	2.8	12
% in green maize	30	1.17	17

Formulation of Concentrate Mixture

Name of the Feed	*Quantity (kg)*	*DM (kg)*	*DCP (kg)*	*TDN (kg)*
Crushed gram	100	90	12	75
Crushed barley	100	90	8	78
Wheat bran	100	90	9	65
Groundnut cake	50	45	21	35
Total	350	315	50	253
Nutritive value (%)	100	90	14	72

Therefore, 1 kg of this mixture contains 0.90 kg DM; 0.14 kg DCP and 0.72 kg TDN.

Composition of Final Ration is

Name of the Feed	*Quantity (kg)*	*DM (kg)*	*DCP (kg)*	*TDN (kg)*
Green berseem	5.00	1.50	0.140	0.60
Green maize	4.00	1.20	0.047	0.68
Wheat straw	5.25	4.725	0.000	2.25
Conc. mixture	3.75	3.375	0.525	2.70
Total	17.75	10.800	0.712	6.23

Provide Mineral and vitamin mixture 30-60 gm/day

Increasing the quantity of green berseem in the ration may reduce the quantity of concentrate mixture.

Example 4

Compute a balanced ration for a pair of heavy working bullocks, each weighing 450 kg. The available feedstuffs with the farmer are wheat straw, green cowpea, crushed gram, barley, wheat bran and mustard cake.

Requirement of one bullock as per standard is:

- ☆ DM (kg): 9.11
- ☆ DCP (kg): 0.64
- ☆ TDN (kg): 5.6

(1) Without Formulating Concentrate Mixture

Name of the Feed	*Quantity (kg)*	*DM (kg)*	*DCP (kg)*	*TDN (kg)*
Green cowpea	7.75	2.32	0.17	0.85
Wheat straw	5.00	4.50	0.00	2.15
Crushed gram	1.00	0.90	0.12	0.75
Crushed barley	1.00	0.90	0.80	0.78
Wheat bran	1.50	1.35	0.13	0.97
Mustard cake	0.50	0.45	0.14	0.39
Total	16.75	10.42	0.64	5.89

Since the above calculation/requirement is of one bullock, hence for the pair the above requirement should be multiplied by 2. So, the total requirement for the pair of bullock will be as follow:

- ☆ Amount of ration 16.75 × 2 =33.50 kg
- ☆ Dry matter 10.42 × 2 = 20.84 kg
- ☆ DCP 0.64 × 2 = 1.28 kg
- ☆ TDN 5.89 × 2 = 11.78 kg
- ☆ Green cowpea 7.75 × 2 = 15.50 kg
- ☆ Wheat straw 5 × 2 = 10.00 kg
- ☆ Concentrate mixture 4 × 2 = 8.00 kg
- ☆ Common salt 30 × 2 = 60.00 gm

The amount of leguminous green cowpea may be increased to reduce the concentrate mixture.

Some Combination of Concentrate Mixture for Cattle/Buffaloes

Feed	Proportion (kg or %)						
	1	2	3	4	5	6	7
Crushed Maize/wheat	20	–	–	20	20	20	10
Crushed Barley	16	–	25	–	20	20	–
Crushed Oat	–	30	–	–	–	–	10
Crushed Guar	–	–	–	–	5	5	–
Crushed Gram	20	20	20	25	5	10	15
Rice bran	–	–	20	25	–	–	15
Wheat bran	30	28	–	–	30	–	15
Mustard cake	–	–	–	–	10	–	20
Til cake	–	–	–	–	5	–	–
Linseed cake	–	15	35	30	5	–	10
Groundnut cake	14	–	–	–	–	5	5
Cotton seed cake	–	–	–	–	–	15	–
Total	100	100	100	100	100	100	100

The above concentrate mixture containing 15-16 per cent DCP and 72-72 per cent TDN.

Depending upon the availability of animal feeds in the locally Dr. S.P. Arora devised some concentrate mixtures for livestock:

Feed	Proportion (kg or %)				
	1	2	3	4	5
Crushed Maize	12	15	20	12	7
Crushed Jowar	10	–	–	–	10
Crushed Bajra	–	–	–	10	–
Crush Barley	–	15	–	–	–
Crushed Oats	–	–	–	–	10
Wheat bran	15	22	10	10	20
Rice bran	15	–	10	20	–
Gram chuni	–	–	10	10	10
Maize gluten	–	10	–	–	–

Feed	*Proportion (kg or %)*				
	1	*2*	*3*	*4*	*5*
Gram husk	5	–	–	–	–
Horse gram	–	–	7	–	5
Sesame meal	10	–	–	–	–
Mustard cake	–	10	–	–	25
Cotton seed meal	–	10	–	10	–
Guar meal	10	–	–	–	5
Linseed meal	10	–	20	–	–
Groundnut meal	10	10	15	20	–
Molasses	–	5	5	5	5
Mineral mixture	3	3	3	3	3
Total	100	100	100	100	100

The above concentrate mixture contains 15-16 per cent DCP and 72-72 per cent TDN.

Feeding of High Producing Dairy Cattle

The yield of crossbreds (above 15 kg milk/day) is many times more than local/desi animals. The rate of milk let down in first six weeks of lactation is so high that the secretion of nutrients into the milk exceeds the rate of uptake of nutrients from the digestive tract. The nutrient deficit is compensated by the diversion of nutrients from the body reserves (mobilization of body protein and fat) resulting in weight loss. Too large a loss in body weight can prove harmful and uneconomical.

The appetite of the animal during the early lactation (up to 8 weeks) is reduced by 2 to 3 kg per day. So all the nutrient needs of the animals are to be provided within the appetite limit. It is difficult to meet the nutrient requirements, particularly the energy requirement of such high yielder (more than 15 kg/d in case of crossbred cows and 12 kg/d in case of buffaloes) through normal concentrate mixtures and fodder. High-energy diets are to be formulated and challenge feeding has to be adopted. And adequate fibre (36 per cent NDF in the total ration) is critical for the maintenance of normal milk fat. Usually, all such cows and buffaloes will remain under negative energy balance during first 5 months of lactation.

Challenge Feeding

It starts 02 weeks before the expected date of calving. Feeding of concentrate mixture should be started initially at 500 gm per day and increase it gradually to a level of 500–1000 gm per 100 kg body weight. High milk producing animals are fed increasing quantity of feed challenging them to produce at their maximum potential. This challenge feeding will condition their digestive system for the increased quantity of feed to provide sufficient nutrients to initiate lactation on a higher plane. This effect has been found to have higher total milk yield in the lactation.

In light of advances made in the field of protein metabolism, the protein requirements in the ruminants are calculated based on rumen protein degradability. Mobilization of body reserves during early lactation can be prevented by feeding high fat, high protein oilseeds such as cottonseed, which supply both protein and long chain fatty acids (LCFAs) for post-ruminal digestion.

Concentrate Mixture Formula for Crossbred and Buffaloes

Ingredients	*Proportion (per cent)*		
	I	*II*	*III*
Maize/wheat/oats/barley	40	30	45
Rice polish	18	20	10
Soybean cake	30	25	35
Groundnut meal	12	15	–
Molasses	07	07	07
Mineral mixture	02	02	02
Common salt	01	01	01
Total	100	100	100

Note: The concentrate mixture should be given to high yielding crossbred lactating cow @ 01 kg/2.5 kg of milk.

Low level of protein and energy in the diets of cows and buffaloes are liable to affect the reproductive system in a number of ways, such as:

1. Disturbing the estrous cyclicity
2. Prolonging postpartum anoestrous period
3. Increasing number of services per conception

In case of energy requirement met fully from time to time during lactation, there may be further increase in milk production as well as better persistency from such cows.

High protein diets are reported to be beneficial for higher milk production and superior growth rate. It has been estimated that two-third of increase in milk yield is due to adequate protein and one-third is a result of optimum energy in the rations.

Feeding of Goat

Goat requires energy in the diet for effective utilization of nutrients and overall productivity. The amount of energy needed increases with activity and distance covered to collect the feed etc. Minerals and vitamin needs are usually met by the forage and grain in the diet, it may be necessary under some conditions to add these nutrients to the ration.

Goat is one of the most economically reared domestic animals. Goat being a small ruminant has a wider range of feed acceptability and crude fiber digestibility than cows and sheep. They eat mainly top feeds, tender buds, twings and grasses, weeds, wild plants and trees, etc.

But goats need considerable care in feeding if they are to produce large quantities of milk or to grow quickly and yield good quality meat. Nutrition and rationing for goats depend on the requirements for maintenance, growth, lactation, foetal development and the daily energy need.

Goats on range or pasture will usually get enough of the necessary vitamins in their diet. The vitamin supplement may be necessary for goats on high producing lactating stage. Antibiotics like auromycin or teramycin can be fed to the young kids. This will increase their growth rate, reduce the incidence of scours and other infectious diseases and improve the general appearance of the kids.

Practical Feeding of Goat

The average milk yield of goat per lactation period of 120 days is about 60 litres, whereas it is about 100 litres in Barberi and 250 litres in Jamunapari. Based on body weight a goat is capable of consuming more feed than a cow (3 to 7 per cent of her live weight in dry matter as compared to 2.25 to 3 per cent in case of cows). Goats differ markedly from sheep and cattle in grazing habits, sensitivity to sweet, salty, bitter and sour taste in accepting or rejecting the feeds. Goats are more tolerant of eating feeds containing bitter principles and refuse any soiled feed. The feed conversion efficiency for milk production of a dairy cow is about 38 per cent , whereas for a goat it ranges between 45 to 70 per cent. Goats are also 4.4 per cent superior to sheep, 7.9 per cent to buffaloes and 8.6 per cent to cows in crude fibre utilization.

A goat uses more feed for its maintenance than a cow and uses more fodder for digestion and metabolism than a cow does.

Feeding of Kids

Kids must receive colostrums from the doe within half an hour after birth and should be continued for three days. For the first seven days the doe with milk and colostrums should nurse the kids. Over-feeding should be avoided and good results are obtained if the frequency of feeding is four or more.

Feeding Schedule for Goat

S.N.	*Age and Stage of Production*	*Feed Ingredients*	*Daily Amount to be Fed*
1.	Birth to 3 days	Colostrum	*Ad libitum*
2.	3 days to 3 weeks	Whole milk or replacer	450 ml
		Water, salt	*Ad libitum*
3.	3 weeks to 4 months (Start minimizing milk & completely stop it when kids attain 4 months of age	Whole milk	400 c.c. up to 8 weeks
		Creep feed	550 g/d
		Lucern hay	*Ad libitum*
		Water and salt	*Ad libitum*
4.	4 months of freshening	Concentrate mixture	15 -16% CP @ 450 g/d
	Dry pregnant	Concentrate mixture	15% CP @400-500g/d
		Lucern hay	*Ad libitum*
		Water and salt	*Ad libitum*
5.	Milking doe	Concentrate mixture	@350 g for each litre of milk
		Trace mineralized salt	1 per cent
		Molasses	5-7 per cent of conc. mixture
6.	Breeding buck	Only pasture	Non-breeding season
		Conc. mixture	@400g/d at breeding season

Creep Ration/Feeding

To obtain a faster body weight gain in kids it is required to provide them a balanced starter feed known as 'Starter ration' upto 10 weeks of age.

Kids should be separated from their dam from 8th day and kept on starter ration till 60th day. They should be allowed restricted suckling 2-3 times a day or should be bottle fed, with fixed allowance. Free choice legume hay, minerals and drinking water should be made available with starter ration.

Composition of Starter Ration

Particulars	*Feeds (per cent)*				
	1	*2*	*3*	*4*	*5*
Crushed maize	35	30	–	30	25
Crush barley/oats	20	30	40	30	40
Crushed maize	35	30	–	30	25
Crush barley/oats	20	30	40	30	40
Wheat bran	12	7	15	7	12
Groundnut cake	20	20	22	10	10
Fish meal	10	10	10	10	10
Molasses	–	–	10	10	–
Mineral mixture	2	2	2	2	2
Common salt	1	1	1	1	1
Calculated value:					
DCP	17	17	17	12	12
TDN	76	75	70	70	72

1. Add 150 g TM-5/Aurofac
2. Add 25 g Vitablend/Rovimix

Feeding Schedule of Kids Upto Weaning Age

Age of Kids	*No. of Feeding/d*	*Dam's Milk or Cow's Milk(ml)*	*Green Feed*	*Starter Feed*
1-7 days	With the dam	Dame's milk	–	–
8-42 days	4	100	*Ad libitum*	Creep feeding
43-60 days	3	100	*Ad lib.* feed	*Ad lib.*

Provide good quality legume hay or fresh green grass and starter ration along with *ad lib* fresh and clean water at 3-4 weeks of age. Equal parts of cracked maize, crushed oats, wheat bran and 10 per cent linseed meal may be fed as the concentrate mixture. Milk replacer may be fed after the kids reach 4 months of age.

Feeding from Four Months to Breeding Age (One year or above)

At this age the kids may be fed roughages that will provide enough nutrients for normal growth. If low quality roughages are fed, supplement the ration with a 12-14 per cent protein ration used for dairy calves @ of 350-400g/d. Do not allow growing dairy goats to become too fat. Reduce the intake of energy feeds as necessary to prevent the fattiness. Always provide fresh and clean water and minerals to kids as they grow.

Grower Feeding Ration

The requirements of DCP and TDN for growing kids after weaning have been given in table. A complete ration providing 9-10 per cent DCP and 60-65 per cent TDN with 20-25 per cent DM from good quality roughage will meet the requirements. If legumes are not available, additional concentrates will have to be provided. The average composition of grower's ration is given in following table.

Composition of Grower Ration

Particulars	*Feed (per cent)*			
	1	*2*	*3*	*4*
Crushed maize	40	10	25	15
Crushed Oats/barley	10	40	30	10
Dal chunies	–	10	20	30
Wheat bran	30	20	10	30
G.N.Cake	10	10	5	5
Molasses	7	7	7	7
Mineral mixture	2	2	2	2
Common salt	1	1	1	1

1. Maize can be replaced with jowar
2. Dal chunies may be replaced by pea, gram and lobia etc.

Finisher Ration

During finishing period, generally 1/3rd gain in body weight is achieved. For good carcass cereal based energy rich feeds are required. The roughage should constitute about 20-25 per cent dry matter for fatty finish and 30 to 40 per cent for lean carcass production. A complete ration providing 5-6 per cent digestible crude protein and 60-65 per cent total digestible nutrients is quite satisfactory for finishing period.

Feeding of Dry Goat

Dry goat should be fed enough to rebuild its lost body reserves. Good leguminous fodders will meet its maintenance and other needs. On pasture, no supplemental feeding is generally required. About six to eight hours of browsing will take care of the normal health. Some common salt and mineral mixture may be needed. Under poor fodder conditions, about 400 grams concentrate mixture in two lots may be fed.

Feeding of Pregnant Goat

High quality roughages provide the basic nutrients needed during the last six weeks of gestation when 70-80 per cent gain in the foetal mass is made. Therefore, liberal feeding of quality leguminous fodder and concentrates having 25 per cent protein should be offered between 400-500 g depending upon the conditions of doe. Free choice of mineral mixture will take care for calcium and phosphorus requirement of dam and foetus. Allow good grazing if available and make sure that does get plenty of exercise. One week before the doe's kidding reduce the amount of concentrate mixture to one half and add bran to provide more bulk. After kidding, feed mash for a few days, gradually bringing the doe to the full feed for more milk production.

Feeding Lactating Goat

The ration of lactating doe should contain high quality roughages like lucern, berseem and other cereal grasses like wheat and paddy straw, through which it will receive not only fresh nutrients particularly of minerals, vitamins and proteins but also the bulk required for volatile fatty acids needed for high milk production. Supplement more nutrients particularly of energy and protein, grain mixture should be given @350 g for each litre of milk or one kg of concentrate mixture over the maintenance ration will be sufficient for every three kg of milk, which may be fed in two lots at the time of morning and evening milking.

Concentrate Mixtures for Lactating Goats

Particulars	*Feed (per cent)*			
	1	*2*	*3*	*4*
Crush maize	37	35	25	30
Crushed Oats	–	37	25	20
Wheat bran	20	16	15	7
Gram	15	–	12	20
G.N.Cake	25	–	20	10
Soybean meal	–	9	–	10
Mineral mixture	2.5	2	2	2
Common salt	0.5	1	1	1

Feeding of Bucks

During the breeding season, the buck should be fed with the same concentrate mixture fed to does at the rate of 450 to 900g daily on the basis of body weight. Provide good quality roughages free of choice along with clean fresh water and minerals. Do not allow the buck to get too fat, reduce the intake of energy feeds.

Non-breeding bucks may normally be fed with 1.35 kg of berseem or other legume hay, 0.5kg of silage and 750 g grain/d. The grain mixture may consist of 100kg crushed maize, 100kg crushed oats, 50 kg of wheat bran and 25 kg linseed meal. During breeding season the grain mixture ration for mature bucks may be increased to about 900g per head per day. When the bucks are rearing on good pasture, no grain mixture feeding is required.

Feeding of Replacement Stock for Breeding

Feeding of female and male kids which are reared for breeding purposes are so adjusted that their sexual maturity and body size are achieved in about one year of age and at 18 to 20 kg live weight in small breeds and 22 to 28 kg in large breeds. After weaning at 8 to10 kg body weight, no supplementary feeding is required on good pasture and browing conditions. However, in lean periods 250 to

400 gm of concentrate mixture per day per head, providing 10 to 12 per cent digestible protein and 68 to 70 per cent total digestible nutrients may be supplied.

Feeding of Sheep

Because sheep are ruminants, a high percentage of their diet comes from roughage. Energy is the most limiting nutrients for production in sheep rations. Energy requirements for nursing ewes are higher than for dry ewes. The quality of protein in the diet does affect the utilization. Methionine is the first limiting amino acid for weight gain and wool growth. Urea can be used in sheep diets, but its use must be limited.

Sheep possesses a unique ability to survive on natural grasses, shrubs and agricultural waste products. Maintaining sheep on concentrate mixture is not the natural method. Pasture and ranges are the natural habitats of the sheep and they thrive on them under the extremely wide varieties of climatic conditions and utilize most adverse types of vegetations.

Inadequate availability of feed and forages due to reduction in area and deterioration of grazing lands poses a serious threat to sheep production because of high livestock density. For this it is essential to develop arid semi waste lands into pastures and provide sufficient top feed through plantation of suitable fodder trees as road side plantation or elsewhere, conserved fodders and to increase production of coarse grains for meeting the requirement of sheep.

Feeding of Pre-weaned Lambs

Under 12 weeks of age lambs suckling the mother should be supplemented with creep mixture.

Composition of Creep Mixture

Particulars	*Feed (per cent)*			
	1	*2*	*3*	*4*
Maize flour	50	37	37	40
Barley flour	14	30	25	27
Wheat bran	10	–	10	10
Soybean meal	6	15	5	–

Particulars	Feed (por cent)			
	1	2	3	4
Groundnut cake	10	5	10	10
Fish meal	7	10	10	10
Mineral mixture	2	2	2	2
Common salt	1	1	1	1
Total	100	100	100	100

To 100 kg of the above feed formulae add 150 g TM-5 and adequate vitablend to take care of vitamin requirements. Feeding of this creep mixture has given a growth rate of 110 -125 g/head/day unto 90 days of age.

Feeding of Growing and Finishing Lambs

When good quality fodder is available then the concentrate requirement is less and when lambs are kept on poor quality fodders the concentrate requirement is more. Following quantities of feeds may be fed to lamb.

Body Weight (kg)	*Good Quality Fodder ad lib + Conc. Mixture (g/d)*	*Poor Quality Roughage ad lib + Conc. Mixture (g/d)*
10-15	50	300
16-25	100	400
26-35	150	600

Composition of Concentrate Mixture

Particulars	Feed (Per cent)	
	1	2
Maize/oats/barley/bajra/wheat/jowar flour	25	35
Wheat bran	40	30
Gram	–	12
Soybean meal/Groundnut cake/Til cake/ Linseed cake/Mustard cake/Cottonseed meal	32	15
Mineral mixture	02	02
Common salt	01	01

TM-5 150 g and vita blend may be added as per requirement.

Growing Ration from Weaning to 6 Months of Age (12-20 kg body weight)

There are three types of rearing system of sheep to be followed.

1. Intensively reared
2. Any one of the followings may be given:
   (*a*) 100g Groundnut cake + dub hay *ad libitum*
   (*b*) 200-250 g concentrates mixture + roughage *ad libitum*
   (*c*) 100 g concentrate mixture + 200 g berseem hay/cowpea + *ad libitum* straw
3. Semi-range conditions
   In addition to usual 8 hours of grazing give any one of the following:
   (*a*) 75 g of groundnut cake
   (*b*) 200 g of concentrate mixture
   (*c*) 150 g of concentrate mixture
   (*d*) 50 gm of concentrate mixture

Particulars	Feeds (Parts)	
	1	2
Crushed Maize	20	–
Crushed Oats	–	31
Wheat bran	60	33
Groundnut cake	20	33

For 15 to 20 kg body weights increase the roughage and concentrates proportionally. Salt and mineral mixture should always be offered free choice. Water supply should also be free choice.

Feeding of Adult Sheep

Free choice maintenance quality fodder like oat, dub grass, maize etc. plus 100 g concentrate mixture may be fed. If leguminous roughages like cow pea, lucern berseem etc are offered in sufficient amounts the feeding of concentrate mixture may be stopped for non-productive stock. Small amount of straw may be provided with such roughages as to prevent digestive disorders. In the absence of good quality fodders, straw may be fed *ad libitum* along with 400g of concentrate mixture.

Feeding of Rams

Generally they are maintained on the same feeding system as ewes. In case they are overfed they should be thinned by gradual reduction in feed and plenty of exercise. During breeding season normal sized rams will require concentrate feeding for a month before as well as during the whole breeding season. At this time an average ram may be offered 250-500 g of concentrate mixture.

1. Rams Reared in Intensive Condition

Feeding *ad libitum* green feed like maize, cowpea, and dub grass would meet the requirement without supplementing concentrates. However, when poor quality roughages are available, concentrates will have to be supplemented to 50 kg body weight 1.5 kg jowar hay and 400g concentrate mixture (contains 18 per cent DCP and 70 per cent TDN).

2. Rams Reared in Semi-range Condition

The roughage part will be taken care by the usual grazing. During lean periods 150 g of concentrate mixture may be fed along with the mineral mixture.

Lactating Sheep

Provide legume hay for first 10 days *ad libitum* and reduce concentrate mixture. On the tenth day of weaning provide 250 g concentrate mixture in addition to *ad lib* legume hay up to 2 ½ months after maintenance allowance is sufficient.

Chapter 2

Evaluating the Effect of Total Mixed Rations on Lactating Cows Under Farmers' Field Conditions

☆ *Kusum Rikhari, Kalpana Joshi, Ashoka Kumar, Ripusudan Kumar and Vir Singh*

ABSTRACT

Effect of three different TMRs having different concentrate composition was observed on nine crossbred lactating cows under farmers' field conditions. Performance in terms of total dry matter intake and milk production and composition in a 90 days' experiment was evaluated. Total dry matter intake was 13.50, 12.12 and 12.27 kg/d and 3.96, 3.34 and 3.38 kg/100kg body wt. for TMR-1, TMR-2 and TMR-3, respectively. Milk production was highest for TMR-1 (8.62 kg), lowest on TMR-2 (7.84 kg) and intermediate for TMR-3 (8.47 kg). Cows on TMR-1 and TMR-3 had higher milk fat and crude protein than cows on TMR-2. Other constituents of milk were higher for TMR-1 and TMR-3 than for TMR-2. Composition of TMR-1 and TMR-

3 resulted in higher milk production, milk constituent percentage compared to TMR-2. TMR-1 resulted in highest total dry matter intake and milk production and composition.

Keywords: *Dry matter intake, Lactating cows, Milk production, SNF, Total solids, Total mixed rations (TMRs).*

Introduction

Feeding concentrate mixture ensures availability of adequate amount of crude protein necessary for sustaining the production potential of animals. Majority of livestock owners are smallholders in India who cannot afford to depend on market for purchase of concentrates. Mere availability of feed resources in rural areas also does not ensure better performance of farm animals. Therefore, the total mixed ration (TMR) feeding system is introduced by the animal nutritionists. The complex process in the body of an animal, in order to translate in to products of human use, such as milk, need to be put in balance, which is possible through the feeding of TMRs. TMR is a mixture of all feedstuffs–forages and concentrates–prepared to contain required nutrients and blended to prevent separation and sorting.

TMR feeding systems deliver balanced nutrition of dairy animals resulting in increased milk production, reduced feed and labour costs, improved butter fat test, reduction in problems relating to metabolic health, etc. (Snowdon, 2003). Feeding mixed rations has been a part of traditional feeding system but it lost much of its value owing to shift of selective feeding of animals in tune with seasonal availability of feeds (Joshi, 2004). The aim of TMRs is to provide sufficient energy and protein to the animal (Hoffman *et al.*, 1998). The type of green roughage (leguminous or non-leguminous) present in TMRs also plays a significant role in the intake and digestibility of TMR (Girdher Naresh and Balaraman, 2003). In dairy cattle, TMR fed cows have shown more milk production in comparison to those, which were fed according to the traditional feeding (Varhegyi *et al.*, 2002). Physiological importance of feeding TMRs has been emphasized as these provide balanced nutrients, optimum digestibility, efficient rumen fermentation conditions and superior nutrient utilization both under field and farm system of management (NATP/ICAR 2004). Maltz *et al.* (1991) reported that daily milk yield, milk fat, protein production and feed conversion ratio (kg feed/kg milk) in goats fed 100 per cent TMR was better than those fed 0, 40, and 60 per cent TMR. Bargo *et al.* (2002) reported Combining pasture and TMR resulted in higher milk production, milk fat and protein percentage and maintenance in body condition score compared to pasture plus concentrate. The TMR feeding system resulted in the highest total dry matter intake and milk production. The present investigation was carried out with three different total mixed rations (TMRs) having different levels of nutrients and their impact on DMI and milk production and composition.

Materials and Methods

Nine crossbred lactating cows were used for 3 months' trial to compare the effects of three different total mixed rations having same forages but different concentrates on animal performance. Cows were selected from the village Jaynagar in Udham Singh Nagar District in Terai area of Uttaranchal. Cows were stratified in three groups of three animals on the basis of milk yield and body weight and randomly assigned to one of three dietary treatments, *viz.*, TMR-1, TMR-2, and TMR-3.

Cows in all the treatments consumed the TMR having the roughage and concentrate in the ratio of 60:40 where the dry and green roughage ratio was also maintained as 60:40 but the composition of concentrates in all the TMRs were different. TMR-1 had mustard oil cake (MOC) in its ingredients which was partially replaced by de-oiled mustard cake in TMR-2 and completely in TMR-3. TMRs were fed *ad lib* to all the animals individually. Clean drinking water was offered twice daily to all animals.

Total dry matter intake (DMI) was measured during last week of experimental period. Cows on all the three treatments were placed in shed during this period and fed individually to measure DMI of the TMR. Total DMI was estimated on the three treatments by conducting digestion trial. During this period daily feed offered and faeces voided were recorded and collected for each animal. Feed samples were analyzed for DM, CP, CF, EE, NFE, NDF, ADF, ADL, Hemicelluloses and cellulose. Fecal samples were thawed, dried at 100°C at hot air oven and ground through 1.0 mm sieve. A composite sample per cow was made for intake measures. The DMI figures for all the treatments were measured as the difference between offered and refused rations. Milk production was recorded daily. Milk samples were collected fortnightly during the experimental period. Individual samples were analyzed for fat, protein, lactose, ash, SNF and total solids.

The data recorded for various parameters were subjected to randomized block design (RBD) as per procedure laid down by Snedecor and Cocharan (1967) to find out the significance of difference among different treatment groups.

Results and Discussion

Chemical Composition

Chemical composition of all feed ingredients of the three TMRs as well as of three mixed rations on dry-matter basis is presented in Tables 2.1 and 2.2.

Table 2.1: Chemical Composition (per cent on DM basis) of Concentrate Mixture and Fodder

Particulars	*Conc. I*	*Conc. II*	*Conc. III*	*Wheat Straw*	*Maize*	*Sorghum*
DM	91.01	91.78	90.43	87.20	20.80	22.00
OM	88.46	87.92	88.94	92.26	92.88	91.48
CP	19.78	18.90	19.25	4.03	8.58	8.40
CF	10.10	10.96	11.20	47.83	36.8	35.40
EE	3.25	2.88	3.50	0.80	1.80	2.20
NFE	55.33	55.18	54.99	39.60	45.70	45.48
NDF	40.72	41.46	40.59	73.18	60.72	65.60
ADF	17.75	18.10	17.88	44.56	33.63	41.50
ADL	3.74	3.77	3.89	9.70	6.80	6.90
Cellulose	14.01	14.33	13.99	34.86	26.83	34.60
Hemicelluloses	22.97	23.36	22.71	28.62	27.09	24.10

Dry matter content of feeds and fodders depends upon the stage of maturity of plant, method of feed processing and preservation, atmospheric moisture etc. The values found in the present analysis would vary or corroborate with other findings depending upon these factors.

Table 2.2: Chemical Composition (per cent on DM basis) of TMRs Fed to Experimental Animals at Farmers' Field

Particulars	*TMR-1*	*TMR-2*	*TMR-3*
DM	38.75	39.16	37.42
OM	90.46	89.98	90.11
CP	10.08	9.97	10.10
CF	33.36	31.79	31.50
EE	2.02	1.95	2.16
NFE	45.00	46.27	46.35
NDF	60.88	60.05	60.10
ADF	33.09	32.58	32.60
ADL	7.80	8.30	8.00
HC	27.79	27.47	27.50
Cellulose	25.29	24.28	24.60

Voluntary Dry Matter Intake

The data obtained for dry matter intake in different TMRs fed to crossbred lactating cows are presented in Table 2.3. The perusal of data revealed that total dry matter intakes in terms of kg/day were non-significant among the treatments. The values of DMI, recorded as 13.50, 12.12 and 12.27 kg/day for TMR-1, TMR-2, and TMR-3 respectively, showed that total DMI was higher for TMR-1 than the rest of the two treatments.

When these values were converted into intake in terms of kg/100kg body weight, the figures were obtained as 3.96, 3.34 and 3.38 kg/100kg body weight for TMR-1, TMR-2, and TMR-3, respectively. The value for TMR-1 showed significantly higher intakes ($P<0.05$) than TMR-2 and TMR-3. The values for TMR-2 and TMR-3 did not vary significantly ($P>0.05$). The values obtained from this study were found on lower side in terms of kg/day and corroborating in terms of kg/100 kg body weight compared to earlier studies. Bargo *et al.*, for example, (2002) reported 24.3 kg/day and 3.73 kg/100kg body weight total DMI for TMR. Schroeder *et al.* (2003) recorded 23.7kg/day and 3.9kg/100kg body weight total DMI for TMR which were also considerably higher in terms of kg/day and similar in terms of kg/100kg body weight compared with the values of the present study.

Table 2.3: Average Dry Matter Intake of Crossbred Cows Fed Different TMRs

Treatments	*DMI (kg/day)*	*DMI (kg/100 kg BW)*
T_1	13.50	3.96
T_2	12.12	3.34
T_3	12.27	3.38
CD at 5 per cent	1.67	0.13
SEM	0.43	0.003

Generally, the intake depends upon the animal species, type of fodder, nutritional status of animal before transfer to the experimental ration and the physiological status of the animal. Values obtained in this study are thus attributable to a combination of such factors.

Milk Production and Milk Composition

The experiment was carried out for 90 days and milk yield was recorded fortnightly and milk samples analyzed at every fortnight. Average milk production and milk composition through all the six fortnights for all the three treatments are presented in Table 2.4.

Milk Production

As explained earlier, the milk yield for each animal was recorded fortnightly. Table 2.4 shows that at the Ist fortnight the milk yields were 7.86, 7.43, and 8.60 kg for TMR-1, TMR-2, and TMR-3, respectively. These values have no significant difference among the treatments. The trend of increase or decrease through all the six fortnights is shown in Figure 2.1.

It is revealed from the Table that an increase was recorded for the treatment 1 till Vth fortnight then it decreased till the end of the experiment. For the treatments 2 and 3, the increase was recorded till IVth fortnight and then it started decreasing till the end of the experiment. The highest milk yield was recorded at the Vth fortnight for the TMR-1 (9.49kg) which was followed by TMR-3 (9.02kg) and TMR-2 (8.38kg) at the end of IVth fortnight. At the end of the experiment, the average milk yield was recorded as 8.62, 7.84, and 8.47 kg for TMR-1, TMR-2, and TMR-3 respectively. These data

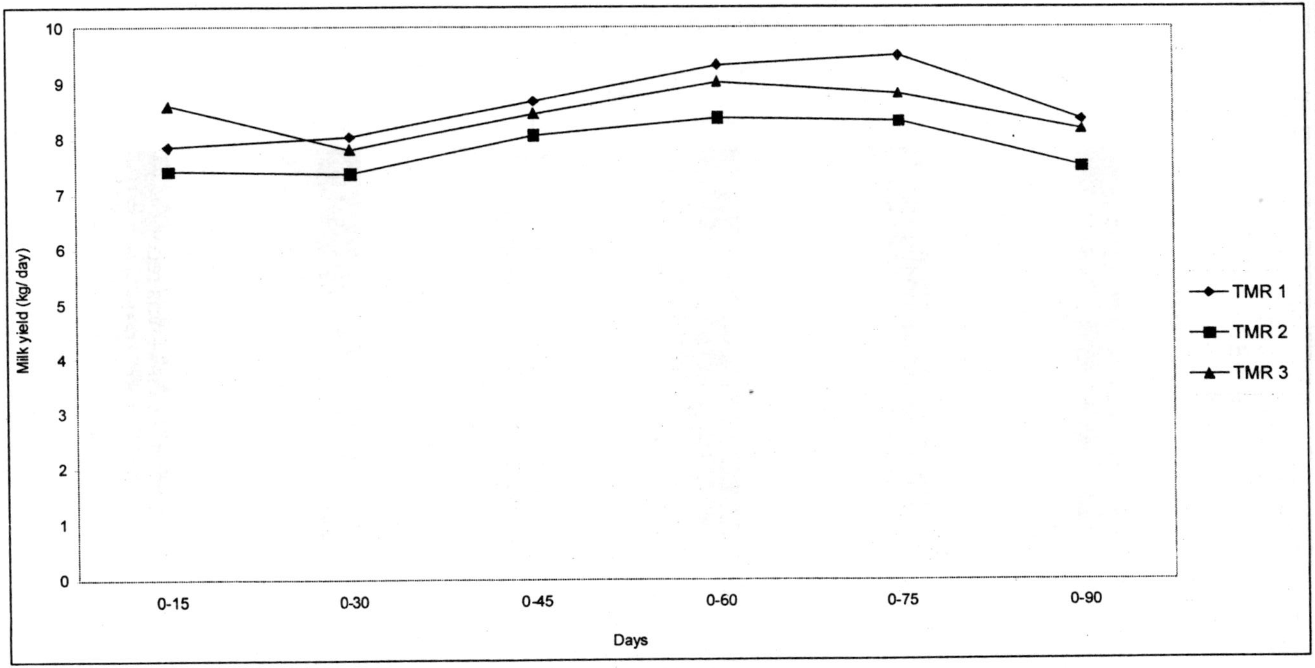

Figure 2.1: Graphical Representation of Milk Yield

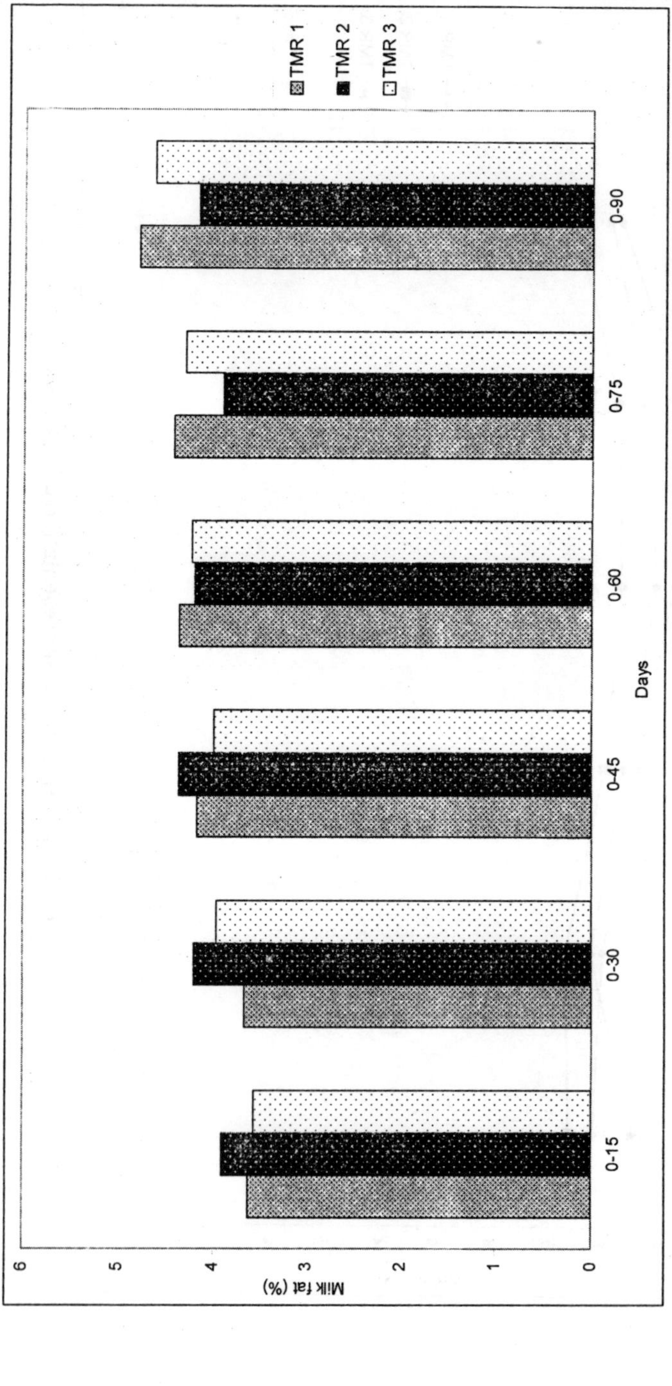

Figure 2.2: Graphical Representation of Milk Fat

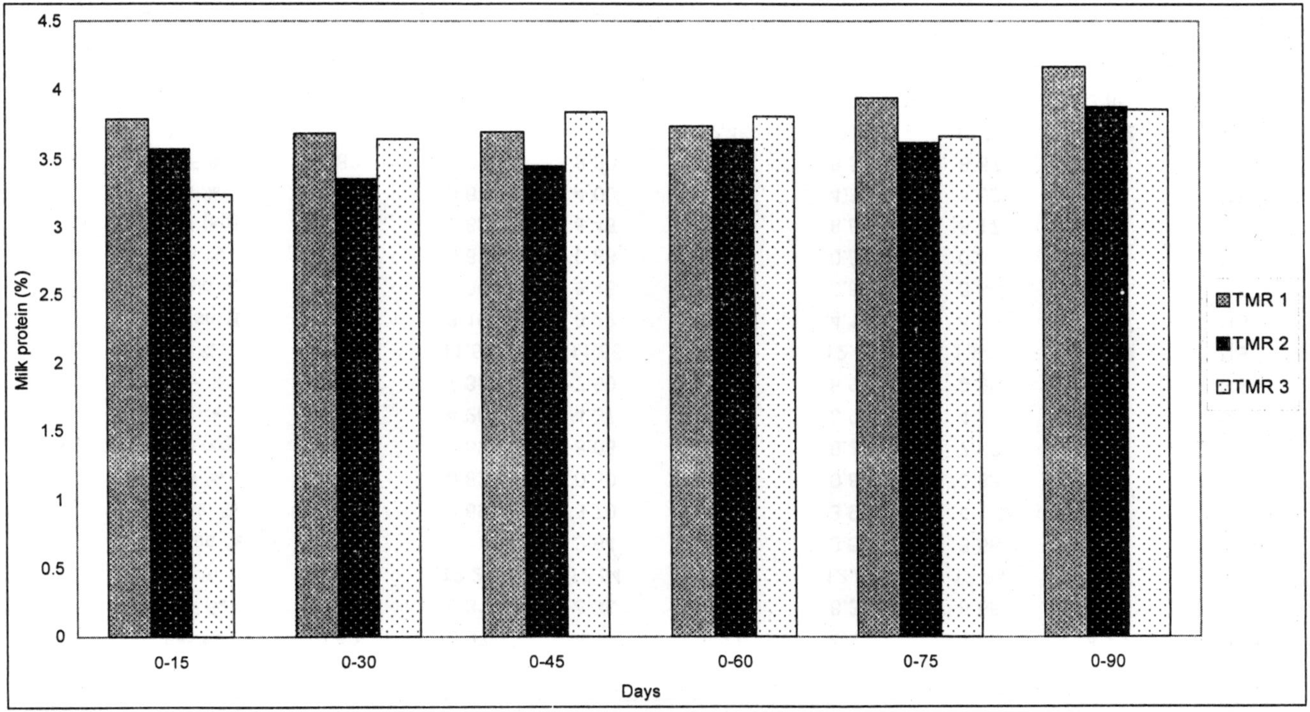

Figure 2.3: Graphical Representation of Milk Protein

Table 2.4: Average Fortnight Milk Yield and Milk Composition in TMRs fed to Crossbred Cows

Treatment	*Particulars*	*I*	*II*	*III*	*IV*	*V*	*VI*	*Avg.*	*CD at 5%*
TMR-1	Yield (kg)	7.86	8.04	8.68	9.33	9.49	8.35	8.62	NS
	Fat %	3.63	3.97	4.17	4.37	4.43	4.80	4.23	NS
	SNF %	8.44	8.31	8.36	8.63	8.33	8.55	8.44	NS
	TS %	12.08	12.28	12.54	12.93	12.87	13.34	12.67	NS
	Lactose %	3.88	3.79	3.94	4.09	3.58	3.56	3.81	NS
	Protein %	3.79	3.69	3.70	3.74	3.95	4.17	3.84	NS
	Ash %	0.77	0.83	0.80	0.80	0.80	0.83	0.81	NS
TMR-2	Yield (kg)	7.43	7.39	8.08	8.38	8.32	7.50	7.84	NS
	Fat %	3.90	4.20	4.37	4.20	3.90	4.17	3.84	NS
	SNF %	7.34	7.36	7.50	8.13	8.97	9.30	8.10	NS
	TS %	11.33	11.56	11.86	12.33	12.54	13.47	12.18	NS
	Lactose %	3.06	3.19	3.22	3.63	4.17	4.56	3.81	NS
	Protein %	3.57	3.36	3.45	3.64	3.62	3.88	3.59	NS
	Ash %	0.80	0.81	0.83	0.86	0.85	0.86	0.83	NS
TMR-3	Yield (kg)	8.60	7.82	8.46	9.02	8.81	8.17	8.47	NS
	Fat %	3.57	3.67	4.00	4.23	4.30	4.63	4.06	NS
	SNF %	7.82	8.21	8.20	8.22	8.54	8.87	8.31	NS
	TS %	11.24	11.95	12.31	12.45	12.90	13.70	12.42	NS
	Lactose %	3.94	3.76	3.61	3.76	4.03	4.33	3.85	NS
	Protein%	3.24	3.65	3.84	3.81	3.67	3.86	3.68	NS
	Ash%	0.80	0.81	0.83	0.83	0.84	0.88	0.83	NS

showed that the average milk yield was highest for TMR-1 which was followed by TMR-3 and TMR-2.

All the values had no significant difference among the treatments. TMR-1 had better performance than the rest two TMRs. These results were found to be similar with those of Soriano *et al.* (2001) who reported that cows in TMR group produced more milk than other treatment groups. Bargo *et al.* (2002) also found higher milk production in cows fed TMR than any other treatment.

Fat Content of the Milk

The fat content in the milk for the three different treatments was recorded throughout the experiment from 1st to 6th fortnight which is briefly described in Table 2.4 and also presented in Figure 2.2. The table reveals that recorded fat percentage at 1st fortnight was similar among all the treatments, *viz.*, 3.63, 3.90, 3.57 per cent for T1, T2, and T3 respectively.

As the experiment went on, the fat percentage increased continuously till the end of the experiment for the TMR-1 and TMR-3 whereas the increase in fat percentage for TMR-2 was not regular. It increased up to 3rd fortnight then decreased till 5th fortnight and again it increased till the end of the experiment. At the end of the experiment recorded fat percentage was 4.8, 4.17, and 4.63 for T1, T2, and T3 respectively.

These values have no significant difference among the treatments. The overall average figure of fat percentage was noted highest in TMR-1 (4.23 per cent) followed by TMR-2 (4.12 per cent) and TMR-3 (4.07 per cent). Though these values have no statistical difference among the treatments however TMR-1 contributes to the highest fat percentage in cow milk. These results indicate better performance of TMR-1 than other two TMRs.

Calberry *et al.* (2003) reported increased milk fat percentage by feeding TMR to lactating cows. The present values are 2.63 per cent higher than those reported by Calberry *et al.* (2003) in their experiment, *i.e.* 2.63 per cent . Soriano *et al.* (2001) reported 3.47 per cent fat in the milk of cows fed TMR which was found to be almost similar to the results of present study.

SNF Content of the Milk

The values of SNF recorded at 1st fortnight were 8.44, 7.34, and

7.82 per cent for TMR-1, TMR-2, and TMR-3 respectively. These values have no significant difference among the treatments. As the experiment was going on the SNF per cent of group 2 and 3 increased till the end of experiment whereas in group 1 the trend of increase of SNF percentage was not regular but at the end of experiment a little increase was seen at the end of the experiment.

The average SNF percentage was recorded as 8.44, 8.10, and 8.31 per cent for treatment 1, 2, and 3 respectively. These values have no significant difference among the treatments but highest value was recorded for treatment 1 which established the superiority for TMR-1 as compared to rest of the two treatments. These values were found to be similar with those of Soriano *et al.* (2001) who reported 8.77 per cent SNF in cows fed TMR.

Protein Content in Milk

The data were recorded fortnightly for protein content in all the three treatments. These data are presented in Table 2.4 and also portrayed in the Figure 2.3. The figures clearly demonstrate the trend of increase/decrease of protein percentage in the milk with the advancement of experiment. At the 1st fortnight the protein percentage was recorded as 3.79, 3.57 and 3.24 per cent for T1, T2, and T3 respectively. These values had no significant difference among the treatments. The average protein percentage in milk was recorded as 3.84, 3.59, and 3.68 for T1, T2, and T3 respectively at the end of the experiment.

These values had no statistical difference among the treatments. Soriano *et al.* (2001) reported 3.23 per cent protein in cow's milk fed TMR which was found to be in the proximity of the figures obtained in the present study.

Lactose Content in Milk

The lactose values showed that at the 1st fortnight the lactose content in milk was almost similar among the treatments, *i.e.*, 3.88, 3.06, and 3.94 per cent for T1, T2, and T3 respectively which did not differ significantly. The lactose percentage started to increase till the 4th fortnight, and then it was decreased till the end of the experiment for T1. For T2 and T3 the lactose percentage increased till the end of the experiment. The average lactose percentage recorded at the end of the experiment was 3.81, 3.81, and 3.85 per cent for T1, T2, and T3 respectively which was almost similar among the treatments without any statistical difference (Table 2.4).

Total Solid Content of Milk

Total solid (T.S.) content of milk presented fortnightly is shown in Table 2.4. The table clearly shows the trend of increase/decrease of T.S. percentage in milk throughout the experiment. At the 1st fortnight the T.S. percentage was found to be 12.08, 11.33, and 11.24 per cent for all the three treatments respectively. At the end of experiment on the 6th fortnight, the T.S. content in milk was recorded as 13.34, 13.47, and 13.70 per cent for T1, T2, and T3 respectively. All the values did not have statistical difference among the treatments. These data show the increasing trend of T.S. in milk with the advancement of the experiment. The average T.S. percentage recorded at the end of the experiment was 12.67 per cent for T1, 12.18 per cent for T2 and 12.42 per cent for T 3 respectively. These values showed that highest T.S. value was recorded for T1 and least value for T2 which established the superiority for T1 as compared to rest two treatments.

Ash Content of the Milk

Ash content of milk was also recorded fortnightly till the end of the experiment (Table 2.4). At the 1st fortnight the ash content was recorded as 0.77, 0.80, and 0.80 per cent for T1, T2, and T3 respectively which at the 6th fortnight was recorded as 0.83, 0.86, and 0.88 per cent for T1, T2, and T3 respectively. These values showed increasing trend of ash content in milk with the advancement of the experiment having no statistical difference among the treatments. The average ash content of milk recorded at the end of the experiment was 0.81, 0.83, and 0.83 per cent for T1, T2, and T3 respectively. These values had no significant difference among the treatments.

Conclusion

Highest numerical value of dry matter intake by the experimental animals was when they were fed TMR-1 and the lowest value was recorded for TMR-2. TMR-1 treatment produced more milk than TMR-3 and TMR-2 treatments. Fat, crude protein, total solids, SNF and lactose percentage in milk was higher in both TMR-1 and TMR-3 treatments than on TMR-2 treatment.

Animal performance was better in TMR-1 treatment compared to TMR-2 and TMR-3 treatments. TMR-3 treatment resulted higher milk fat and protein percentage than TMR-2 treatment.

Acknowledgements

This chapter is an extract of the thesis research of the first author. Grants from NATP (PSR-34) and facilities provided by the College of Veterinary & Animal Sciences of the GB Pant University of Agriculture & Technology are gratefully acknowledged. Farmers' cooperation and full participation in the study village is heartily appreciated.

References

Bargo, F., L.D. Muller, J.E. Delahoy and T.W.Cassidy, 2002. Performance of high producing dairy cows with three different feeding systems combining pasture or total mixed rations. *Journal of Dairy Science*. 85: 2959-2974.

Calberry, J.M.; J.C. Plaizier; M.S. Einarson; and B.W. McBride, 2003. Effects of replacing chopped Alfalfa hay with alfalfa silage in a total mixed ration on production and rumen conditions of lactating dairy cows. *Journal of Dairy Science*. 86: 3611-3619.

Girdhar Naresh and N. Balaraman, 2003. *In vitro* evaluation of maize fodder based total mixed rations containing different levels of energy and protein. *Indian Journal of Animal Sciences*. 73 (11): 1263-1266.

Hoffmann, U.; H. Steingass; R. Funk; Schmettow, R. V.; and Drochner, W.,1998. Impact of total mixed ration. *Kraftfutter*, 7-8: 292-303.

Joshi, K., 2004. *Nutritive Value and In Sacco Degradation Characteristics of Concentrates with Green Maize and Wheat Straw Based Diets*. M.Sc. Thesis. Pantnagar: GB Pant University of Agriculture and Technology.

Maltz, E., N. Silanikove, Y. Karaso, G. Shefet, A. Meltzer and M. Barak, 1991. A note on the effects of feeding total mixed ration on performance of dairy goats in late lactation. *Anim. Feed Sci. Tech.*, 35 (1-2):15-20, 11 ref.

NATP/ICAR, 2004. *Evaluation of Locally Available Feeds and Fodders, Improve Quality and to Formulate Economic Rations in Rice-Wheat*

Based Production System (PSR-34): Progress Report. GB Pant University of Agriculture and Technology, Pantnagar, India, pp 120

Schroeder, G.F.; J.E. Delahoy; I. Vidasurreta; F. Bargo; G.A. Gagliostro; and L.D. Muller, 2003. Milk fatty-acid composition of cows fed a total mixed ration or pasture plus concentrates replacing corn with fat. *Journal of Dairy Science*. 86: 3237-3248.

Snedecor, G.M. and Cochran, W.G. 1967. *Statistical Methods*. New Delhi: Oxford & IBH.

Snowdon, M., 2003. *Total mixed rations for dairy cattle*. Patz Sales, Inc., USA, 104

Soriano, F.D.; C. E. Polan; and C.N. Miller, 2001. Supplementing pasture to lactating Holsteins fed a total mixed ration diet. *Journal of Dairy Science*. 84: 2460-2468.

Varhegyi, J., Z. Juhasz and K. Nemeth, 2002. The effect of feeding by-pass protein during the last two weeks of pregnancy on milk production in the subsequent lactation. *Allattenzyeszts-es-Takarmanyozas*. 51 (2): 134.

Chapter 3

Evaluation of Maize and Sorghum Fodders for Nutritive Value at Two Different Stages of Growth

☆ *Woda Jeremaih Odok, Mehandra Singh, Vir Singh, Ashoka Kumar and D. Kumar*

ABSTRACT

Maize and sorghum fodders at two different stages of growth were fed to sixteen growing crossbred heifers for their nutritive evaluation. The overall DMI (kg/d) of maize (3.02) was significantly ($P < 0.01$) higher than that of sorghum (2.63). The voluntary intake of DMI (kg/d) of maize fodders at stage first (3.05) was also significantly higher than at stage second (2.59). The average content of DM, OM, ADF, Cellulose, NDF, ADL, hemicellulose, and GE were higher in maize than in sorghum. While the DM, OM, NDF, hemicellulose, ADF, cellulose, ADL and GE contents increased, the CP and total ash content decreased from first stage to second stage in these two green fodders. The two species of fodder crops grown differed in their nutritive value under similar agronomic practices.

Keyword: *Maize, Sorghum, Nutritive value, Chemical composition.*

Introduction

There are several factors that usually affect forage quality for instance species, variety, stage of growth, fertility of soil etc. The stage of maturity of forage at the time of harvesting has marked influence on their chemical composition, feeding value and digestibility (Gupta, 1969) whereas harvesting at very early stage to get better nutritive value has adverse effect on dry matter yield. Moreover, delay in harvesting yields in higher dry matter and poor nutritional value. Therefore, an ideal approach to harvest the forage at the stage when the digestible nutrients per unit weight are maximally available to the animals is needed. This study was carried out to evaluate the nutritive value of Maize and Sorghum at different stages of growth.

Materials and Methods

Sixteen growing crossbred heifers aged between 12.21 to 13.375 months and 90.75 to 102.63 kg body weights were randomly divided into two homogenous groups of 8 animals each and allotted to one of the two fodders at two stage of growth. The experiment was carried out in two periods and the fodders used in period I included Maize and Sorghum at stage I. In Period II the same fodders were used but harvested at stage II. During the last 7 days of each experimental period a digestion trail was conducted to evaluate the availability of nutrients and their digestibility. Feed and faecal samples were analyzed for proximate principles of DM, CP (AOAC, 1975) and cellwall constituents of NDF, ADF, hemicellulose, cellulose (Goering and Van Soest, 1970) and energy (O'Shea and Maguire, 1962). The data were analyzed statistically using 2 factors RBD as per Snedecor and Cochran (1967).

Results and Discussion

The various chemical and cell wall constituents of the two green fodders *viz.* Sorghum and maize at two different stages of growth studied are shown in Table 3.1. From the table it seems that the two stages of growth yielded almost similar results for the two crops keeping in view the stage of harvesting relatively had higher DM. This is in agreement with Chauhan *et al.* (1976) who reported that the dry matter content was higher in maize and sorghum than in millet and hybrid nappier NB- 21. The variation of the DM between two crops might be due to variation in genetic potential between the species and probably due to variation in days of harvest at time of evaluation. The content of DM, OM, ADF, Cellulose, NDF, ADL, hemicellulose, and GE were higher in maize than in sorghum. While the DM, OM, NDF, hemicellulose, ADF, cellulose, ADL and GE contents increased, the CP and total ash content decreased from first stage to second stage in these two green fodders. The DM digestibility coefficients of nutrients are shown in Table 3.2. It is evident from the table that the DM digestibility of the two fodders at stage first (70.34 per cent) was significantly ($P < 0.01$) higher than the dry matter digestibility of the two fodders at second stage (61.86 per cent). The two fodders did differ significantly ($P< 0.01$) in their DM digestibility. The overall digestibility of maize was higher than that of sorghum. The CP digestibility of maize was significantly higher ($P< 0.01$) than that of sorghum. This variation was due to variation in chemical composition and palatability. These results disagreed with those reported by Randhawa *et al.* (1988) and Chauhan *et al.* (1976) who reported that sorghum had more CP digestibility than maize.

Table 3.1: Chemical Composition and Energy Value of Maize and Sorghum at Two Different Stages of Growth (per cent DM basis)

Parameter	*Maize*		*Sorghum*	
	Stage I	*Stage II*	*Stage I*	*Stage II*
Dry matter	18.00	28.77	18.33	29.80
Organic matter	89.80	91.80	90.48	91.60
Crude protein	11.03	8.05	11.65	7.18
NDF	60.50	66.00	57.00	65.60
ADF	33.50	38.00	37.00	41.50
Hemicellulose	27.00	28.00	19.30	24.10
Cellulose	27.90	31.30	31.90	34.60
ADL	5.60	6.70	5.80	6.90
Energy	10.20	8.20	9.52	8.40

Table 3.2: Average per cent Digestibility Coefficient of Nutrients and Voluntary Intake of Dry Matter

Treatments		*DM*	*OM*	*CP*	*NDF*	*ADF*	*Hemi-cellulose*	*Cellulose*	*Gross Energy*	*DMI (kg/day)*	*DMI (kg/100 kg bwt)*	*DMI (g/kgw$^{0.75}$)*
T_1 (Rio Stage–I *i.e.* 70–77 DAS)		67.36	69.22	68.27	62.73	58.43	73.23	54.91	67.36	2.895	2.98	92.53
T_2 (Rio Stage–II *i.e.* 100–107 DAS)		58.08	61.44	59.80	56.08	51.62	67.50	48.83	58.14	2.380	2.30	73.19
T_3 (Maize Stage–I *i.e.* 60–67DAS)		73.32	74.46	73.37	69.76	61.73	82.97	59.17	73.18	3.230	3.64	110.98
T_4 (Maize Stage–II *i.e.* 90–97DAS)		65.63	68.55	67.42	61.73	56.25	71.41	53.68	65.57	2.795	3.10	98.25
Fodder	Sorghum (avg.)	62.72	65.33	64.01	59.41	55.03	70.37	51.87	62.75	2.630	2.64	82.86
	Maize (avg.)	69.48	71.51	70.40	65.75	58.99	77.19	56.43	69.38	3.020	3.37	104.61
Stage	Stage–I (T_1+T_3) avg.	70.34	71.84	70.82	66.25	60.08	78.10	57.04	70.27	3.050	3.31	101.75
	Stage–II (T_2+T_4) avg.	61.86	65.00	63.61	58.91	53.94	69.46	51.26	61.86	2.590	2.70	85.72

Note: Values bearing different superscript in the same column indicate significance at ($P< 0.01$) and ($P<0.05$).

DAS: Days after sowing.

The per cent OM digestibility of sorghum and maize at first stage of harvest was significantly ($P < 0.01$) higher than that of at second stage. In both fodder species the OMD decreased with advanced in stages of maturity. Kang *et al.* (1987) reported higher OM digestibility in both crops than in present result. The over all OM digestibility of maize was higher than sorghum. The CP digestibility of maize was significantly higher ($P< 0.01$) than that of sorghum. This variation was due to variation in chemical composition and palatability. These results disagreed with that reported by Randhawa *et al.* (1988) and Chuhan *et al.* (1976) who reported that sorghum has more CP digestibility than maize. The difference in digestibility coefficient of NDF among different treatments was significantly higher ($P< 0.01$). The NDF of sorghum at first stage (62.73) was significantly higher than at second stage (56.08). The significant decrease in NDF digestibility from first stage to second stage might be due to more structural carbohydrates synthesis and lignin deposition in the cellwall during later stages of growth. The results of first stage is in line with that reported by Ripusudan (2001) who reported 64.05 and 66.80at first and second stage of harvest, respectively. In case of maize the NDF digestibility did differ significantly between two stages of growth (69.76 vs 61.73). However, the NDF digestibility in first stage was significantly higher than the second stage indicating that the digestibility of NDF is influenced by the stage of growth.

The ADF digestibility of sorghum at first stage (58.43) was found to be significantly higher ($P< 0.01$) than the second stage (51.62). These results are higher compared to findings of Pradhan *et al.* (1991). In case of maize the ADF digestibility of 61.73 per cent at stage first was higher than stage second (51.62) but not significantly different ($P < 0.05$). The ADF digestibility of maize at two stages of growth was higher than that of sorghum. However, these results were contrary to those of Badwaik *et al.* (1998) who reported higher ADF digestibility in sorghum than in maize.

The hemicellulose digestibility in maize was higher than in sorghum. Such variation among the two crops might be due to genetic difference leading to more deposition of hemicellulose in maize (Table 3.2), digestibility and palatability of the crops. Non significant ($P< 0.01$) difference in the digestibility of gross energy between to two crops at stage first and highly significant ($P< 0.01$)

difference at stage second was observed. However these results disagree with those reported by Randhawa *et al.* (1988) who reported significantly higher value of sorghum at stage first than stage second for both the crops.

The mean value of total dry matter intake (kg/d) was significantly different ($P < 0.01$) amongst the treatments. The dry matter intake (kg/d) for sorghum at first stage (2.895) was significantly ($P < 0.01$) higher than at second stage (2.380). This variation in DM I that is expressed as kg/d might be due to lesser NDF content at stage first leading to better palatability of the fodder at younger stage of harvest. Higher dry matter intake (kg/d) in maize at stage first (3.230) than at second stage of growth (2.795) was observed perhaps due to same reasons and palatability of maize fodder at younger stage.

The mean intake value of dry matter (kg/100 Kg bwt) in maize at first (3.64) and second (3.10) stage was significantly ($P < 0.01$) higher than sorghum at first (2.98) and second (2.30) stage respectively. The maize crop was younger than sorghum and its highly digestible hemicellulose. The mean dry matter intake (kg/100 kgbwt) fed on sorghum at stage first was significantly ($P < 0.01$) higher than that fed on the same fodder at second stage. This variation in dry matter intake (kg/100 Kg wt) at two different stages of growth followed similar trends as in case of DMI (kg/d). The DMI (kg/100 body wt) of heifers fed on maize at first stage (3.64) was significantly ($P < 0.01$) higher than those group fed on sorghum at first and second stage of growth. The present findings of study are in agreement with values reported by Ripusudan (2001) and Chauhan *et al.* (1984). The DMI (kg/100 kg wt) of heifer fed on maize at I stage was significantly higher than that group fed on maize at II stage. Similar findings were reported by Chauhan (1980).

The dry matter intake of (g/kg w 0.75) of group of heifer fed on maize stage first (110.98) and maize at stage second (98.25) were significantly ($P < 0.01$) higher than those of group fed on sorghum at stage first (92.53) and sorghum at stage second (73.19).

These results are in agreement with those reported by Ripusudan, (2001) and Sanjeev *et al.* (1997). The over all Dry matter intake of maize in terms of (kg/day), (kg/100 bwt) and (g/kg $w^{0.75}$) was significantly higher than of sorghum because maize has more palatability and is more soluble fraction (Table 3.2) and DAS were also iess. From the results it is evident that the two species of fodder crops grown differed in their nutritive value under similar agronomic practices. Therefore, it may be concluded that fodders be harvested at younger stages to have better digestibility and intake of nutrients.

Acknowledgement

The financial help received from the Indian Council of Agricultural Research, New Delhi through an *ad hoc* project and Indian Council for Cultural Relations (ICCR) are acknowledged.

References

AOAC. 1975. Official methods of Analysis 11th edn. Association of Orricial agricultural Chemists. Washington, D.C.

Chauhan, T.R. 1980. Note on Metabolizable energy contents of maize (*Zea mays*) fodder at its different stage of maturity. *Indian J. Animal Sci.*, 50 (12): 1134–1136.

Chauhan, T.R.; Gill, R.S. and Ichhponan, J.S. 1976. Metabolizable energy content of some kharif fodders when fed to buffaloes. *Indian J. Anim. Sci.*, 46(10): 520–524.

Chauhan, T.R.; Sidhu, B.S. and Chora, A.K. 1984. Comparative nutritive value of NB–21 and some promising strains of pearl millet Napier hybrid for buffaloes. *Indian J. Anim. Sci.*, 54(1): 1031–1034.

Georing, H.K. and Van Soest, P.J. 1970. Forage fibre analysis (apparatus, reagents, procedure and applications) Agricultural Hand Book No. 379: 1–12 Agriculture Research Services, United States. Department of Agriculture.

Gupta,U.P. 1969. Genetic evaluation of grain and fodder quality of pennistuma; Final research report. Pl. 430 Project A7–CR. 135. U.S.D.a., PAU. Ludhiana.

O'Shea, J. and Maguire, M.F. 1962. Determination of calorific value of feed stuffs by chromic oxide oxidation. *J. Food Sci. and Agric.*, 13: 530–534.

Pradhan, K.; Bhatia, S.K. and Sangwan, D.C. 1991. Relative rumen ecosystem and nutrient digestibility in cattle and buffalo fed high fibrous diets. *J. Res. Haryana Agril. Univ., Hissar,* 103p.

Randhawa, S.A.; Gill, R.S.; Gill, S.S. and Hundal, L.S. 1988. Effect of feeding green sorghum its silage or hay on milk production in buffaloes. *Indian J. Dairy Sci.*, 41(3): 255–257.

Ripsudan, K. 2001. Evaluation of sorghum, millet and maize for their nutritive value at two stages of maturity. Thesis, Master of science in Agriculture (Animal Nutrition). Govind Ballabh Pant University of Agriculture and Technology, Pantnagar.

Snedecor, G.W. and Cochran, W.G. 1967. Statistical methods. 6th ed. Iowa State University Press, Ames. 593p.

Sanjiv, K., Garg, M.C. and Kumar, S. 1997. Nutritional evaluation of M.P. Chari (*Sorghum bicolor*) forage in Murrah heifers. *Indian J. Anim. Nutr.*, 14(4): 281–282.

Chapter 4

Livestock Contributions to Food Security in Mountain Areas

☆ *Vir Singh*

ABSTRACT

Livestock are an integral component of a mountain farming system. Mountain region, in fact, is a natural animal gene bank. Cattle, buffaloes, sheep, goats, carry pack animals, poultry birds, and Angora rabbits occupy a long range of environments, while yaks and bison are concentrated in upper reaches of the Himalayas. Because of the unique topographic and other specific characteristics of mountain areas, farming communities cannot do without an overwhelming involvement of livestock in farming systems. Their livelihoods are intricately woven around livestock resources. Livestock contribution to mountain and highland areas is much more crucial than imagined by planners and policy makers.

The India vs world picture of livestock genetic resources reveals that Indian environments harbour high degree of diversity of livestock genetic resources. The biomass transfer and cyclic flow pattern of nutrients mediated by livestock infuse vitality in the production system. Organic linkages amongst forest ecosystems, livestock and agro-ecosystems is vital for sustainability of mountain agriculture. Livestock thus are central to a mountain farming system. Livestock in mountain farming systems perform several functions which range from agricultural operations, like ploughing, puddling, levelling, interculture and earthing up, threshing, etc. to direct, visible contributions in terms of supplying physical items like dung and milk, to less visible gains in terms of, *e.g.*, employment, income generation, farmers' security and companionship, sustaining livelihoods, sustainability of the farming system etc. Role of livestock in terms of their asset value, output value, employment generation, energy provision and organic farming deserves appreciation and needs to be well thought out planning for the development of this sector.

An improvement in livestock sector through effective local level planning with focus on the conservation of biodiversity comprising unique livestock species and genetic resources would lead to an enormous improvement in the socioeconomic condition of local people. Local availability of feeds, livestock genetic resources, health cover, marketing facilities, etc are the most important factors on which local level livestock development planning can be based.

Keywords: *Asset value, Food security, Livestock genetic resources, Mountain farming systems.*

Introduction

India has an enormous wealth of livestock, especially the dairy animals. There are some 319 million cattle, 94 million buffaloes and 123.50 million goats in the country which constitute about 16 per cent, 57 per cent and 17 per cent of the world's total dairy animal population, respectively (FAO 2001). Livestock again comprise one of the most important sectors of the agrarian economy of mountain areas throughout the length and width of the mountains and highlands of the world. The mountain habitats, distinguishable from the mainstream plains, are inhabited largely by livestock-dependent farming communities. Mountain economies, often decentralized and cash-starved, can boast of their livestock resources–large number of species of bovine, ovine, and non-ruminant animals and their unique genetic resources. Mountain region, in fact, is a natural animal gene bank. Cattle, buffaloes, sheep, goats, carry pack animals, poultry birds, and Angora rabbits occupy a long range of environments, while yaks and bison are concentrated in upper reaches of the Himalayas. Because of the unique topographic and other specific characteristics of mountain areas, farming communities cannot do without an overwhelming involvement of livestock in farming systems. Their livelihoods are intricately woven around livestock resources. Livestock contribution to mountain and highland areas is much more crucial than imagined by planners and policy makers.

Livestock again are the important sector that demands well thought out and well-planned institutional intervention for the socioeconomic development of mountain areas. This sector, however, is the most neglected one in institutional policies and planning. Furthermore, conventional planning, a replica of the one being implemented in the mainstream plain areas, is less conducive to specific mountain conditions. Such a planning not only leads to huge wastage of money and resources but also seeks no participation of local people and ignores all local realities.

Livestock Genetic Diversity in India

There are some 61well described breeds of cattle in the world. India shares about 50 per cent of the total cattle breeds of the world. India has 10 described breeds of buffaloes which again constitute about 50 per cent of the world's buffalo breeds numbering 19. A total of 42 breeds of sheep are found in India out of a total of 59 in the world, *i.e.*, as many as 71 per cent of the sheep breeds of the world are found in India. A total of 20 out of a total of 29 breeds of goats, or about 69 per cent, are in India. Share of the world's breeds of horses, camel and poultry in India is also remarkable (Table 4.1).

Table 4.1: Animal Genetic Resources in India and in the World

Animal	*Number of Breeds*	
	World	*India*
Cattle	61	30
Buffaloes	19	10
Sheep	59	42
Goats	29	20
Horses	9	6
Poultry	18	18
Camels	–	8
Pigs	3	–
Donkeys	3	–

Source: Gautam (2004).

Table 4.2: Attributes of Livestock Farming and their Implication to Mountain Specificities

	Attributes of Livestock Farming	*Mountain Specificities*					
		I	*F*	*M*	*D*	*N*	*A*
I	Physical/Biological						
	Physical						
(*i*)	Mobility	x					
(*ii*)	Sturdiness (adaptability) etc.	x	x	x	x		
	Biological						
(*i*)	User of crop by-products and other diverse biomass materials	x			x		
(*ii*)	Compatible with land extensive resource use system (*e.g.* grass, tree fodder, bush etc.)		x		x		
(*iii*)	User of spatial and temporal variable resources	x			x		x
(*iv*)	Contributor to local resource supply (*e.g.* energy-manure, draft power and food)	x			x		x
(*v*)	Agency for input recycling (*e.g.* bedding materials into manures:						
	– fodders into manure						
	– fodders into firewood)	x			x		
II	Socioeconomic						
(*i*)	Producer of high value, low weight products (*e.g.* ghee, wool etc.)	x				x	
(*ii*)	Producer of locally available inputs (*e.g.* draft power, manure etc.)	x			x		
(*iii*)	Neutral to economies of scale (*e.g.* feasibility of small/large scale of operation)			x	x		

Contd...

Table 4.2–Contd...

Attributes of Livestock Farming	*Mountain Specificities*					
	I	*F*	*M*	*D*	*N*	*A*
(*iv*) Possibility of livestock farming without private land resource base			x	x	x	
(*v*) Reducer of rural inequalities			x			
(*vi*) Conducive to release local pressure	x		x		x	
(*vii*) Generator of employment and livestock outputs all year round						x
(*viii*) Major source of agriculture diversification (due to non-covariate flow of output, employment etc. Compared to crop production system)						
– crop-livestock mixed farming		x		x	x	
– mixed composition of livestock holding				x		
– cushion for handling crisis situations	x		x			
– seasonally differentiated pattern of production flow			x			
(*ix*) Conducive to self-help activities				x		
– sharing of bullock power						
– exchange of livestock products with						
Non-livestock commodities			x			

Contd...

Table 4.2–Contd...

Attributes of Livestock Farming	*Mountain Specificities*					
	I	*F*	*M*	*D*	*N*	*A*
(*x*) Generator of provision of linkages						
(*a*) Inter-activity linkages (*e.g.* crop-livestock-forestry)		x		x	x	
(*b*) Market linkages						
– use of animals for transportation purposes (*e.g.* sheep, goat, yak, chauri)	x					
– sheep move, with commodity (*e.g.* wool)	x					
– petty trading (*e.g.* high value, low weight product)	x					
(*c*) Upland-lowland linkage/complimentarily						
– transhumance	x	x		x		
– means of transport	x					

Source: Jodha and Shrestha (1990).

Note: I: Inaccessibility; F: Fragility; M: Marginality; D: Diversity; N: Niche; A: Human Adaptation Mechanism.

Table 4.3: Contribution of Different Livestock Species to Mountain Farming Community in Uttarakhand

Contributions	*Cattle*	*Buffaloes*	*Goats*	*Sheep Animals*	*Pack*	
	Male	*Female*				
AGRICULTURAL OPERATIONS						
Ploughing	x					
Levelling	x					
Puddling	x					
Weeding, Earthing-up	x					
Threshing	x	x				
Loading, Pack-carrying	x[a]					x
PHYSICAL PRODUCTS						
Dung, Manure	x	x	x	x	x	x
Milk		x	x			
Meat				x	x	
Wool					x	
INCOME/EMPLOYMENT GAINS						
Direct Productivity Improvement	x					
Smaller Gains through Sale	x			x	x	
Larger Gains through Sale	x	x	x			x
Off-farm Activities	x					x
Income through Hiring-out	x					
SOCIAL, CULTURAL, ECOLOGICAL GAINS						
Cropping Diversification	x					
In-situ Manuring of Fields	x	x		x	x	
Renewable Energy Supply	x					
Religious, Ethical, Aesthetic Values	x	x				
Festivity, Fairs, Rituals	x	x				
Social Status, Prestige	x	x				
Social Cohesion Encouragement	x	x				
Farming System Sustainability Enhancement	x					

Source: Singh (1998).

x[a]: in only transhumant societies.

It is revealed from this India vs world picture of livestock genetic resources that Indian environments harbour high degree of diversity of livestock genetic resources.

Many of the Indian breeds have unique traits. For instance, Sahiwal, Red Sindhi, Tharparkar and Gir breeds of Indian cattle, Bhadawari and Toda of buffaloes, Hissardale, Niligiri, Changtangi, Muzaffarnagri, Pugal and Garole of Sheep, Jamanapari, Beetal and Changthangi of goats, Aseel, Kadaknath and Naked Neck of poultry and Zanskari and Spiti of pack animals depict unique and some superb traits. These unique Indian breeds, however, are now showing declining trends (Gautam 2004).

Livestock-centred Farming Systems: Basis for Sustainable Food Security

Livestock acquire special importance in mountain farming systems both on ecological and socio-economical grounds. They are an integral part of a farming system and a lively 'bridge' connecting two types of lands/ecosystems, *viz.*, the uncultivated forestland/forest ecosystem and the cultivated land/agro-ecosystem. This linkage is crucial for the ecological and economical sustainability of the system. In the mountains, as in many parts of the world, the productivity of the farming systems pivots on animals' ability to convert fodder into manure. In mountain areas, especially on high altitudes, crop residues decompose very slowly. Ruminant digestive system helps speed up nutrient recycling in the cropland. Ruminants also help transfer forest soil nutrients available in the forest vegetation into the cropland, further improving the fertility status of the agro-ecosystem.

The biomass transfer and cyclic flow pattern of nutrients mediated by livestock infuse vitality in the production system and livestock themselves fulfill their requirements for maintenance and production. This dynamic relationship among forest ecosystems, livestock and agro-ecosystems is vital for sustainability of mountain agriculture. Livestock thus are central to a mountain farming system. They have very high compatibility with mountain specificities (Table 4.2) and a special focus on them would contribute to sustainability in mountain farming systems.

Livestock in mountain farming systems perform several functions which range from agricultural operations, like ploughing, puddling, levelling, interculture and earthing up, threshing, etc. to direct, visible contributions in terms of supplying physical items like dung and milk, to less visible gains in terms of, *e.g.*, employment, income generation, farmers' security and companionship, sustaining livelihoods, sustainability of the farming system etc. (Table 4.2).

The agrarian economy in mountain areas thus revolves round the livestock. Tulachan *et al.* (2000) have presented a very comprehensive account of livestock contribution to livelihoods in mountain and highland ecosystems of the world. The biomass transfer and cyclic flow pattern of nutrients mediated by livestock infuse vitality in the production system and livestock themselves fulfill their requirements for maintenance and production. This dynamic relationship among forest ecosystems, livestock and agro-ecosystems is vital for sustainability of mountain agriculture. Livestock thus are central to a mountain farming system. They have very high compatibility with mountain specificities (Table 4.2) and a special focus on them would contribute to sustainability in mountain farming systems.

Livestock in the Mountains of Uttarakhand

Uttarakhand, the mountainous State of India, has a variety of livestock–cattle, buffaloes, goats, sheep, horses, ponies, mules, and pigs–numbering 4608656, according to the 1997 Livestock Population (Table 4.4). There are 955012 poultry fowl and 17257 other poultry birds in the State. Among these animals, cattle constitute the highest percentage of population (43 per cent), followed by buffaloes (24 per cent), goats (23 per cent), sheep (7 per cent), and horses, mules, pigs and others (1 per cent each) (Figure 4.1).

Livestock species are specific for specific role and function. For example, cattle are mainly used as milch and draught animals, while buffaloes are exclusively dairy animals. Ponies, mules and horses are used as carry pack animals. Sheep and goats are meat animals, while sheep are reared primarily for wool production. Angora rabbits are for fibre. Poultry birds are for meat and eggs. Populations of livestock are distributed according to geo-ecological zones. While lower tracts of the mountains harbour cattle, buffaloes, goats and pack animals, the upper areas are more suitable for sheep rearing.

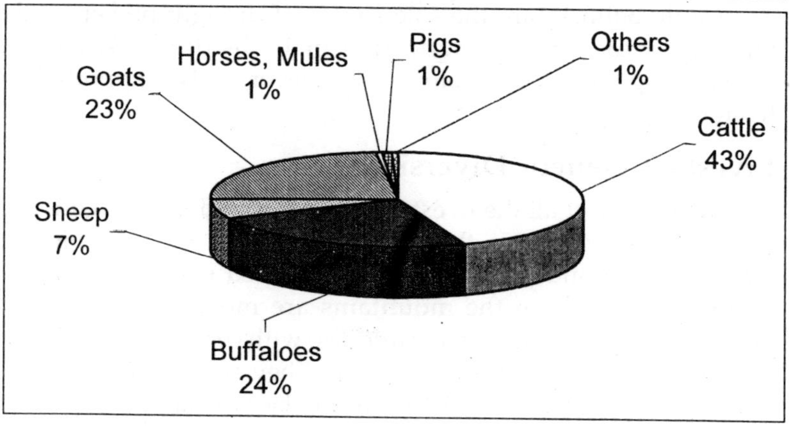

Figure 4.1: Livestock Composition in Uttaranchal

Table 4.4: Population of Different Species of Livestock in Uttarakhand

Cattle	2030856
Buffaloes	1094295
Sheep	310750
Goats	1085700
Horses, Mules	23560
Pigs	31551
Others	31944

Source: Kurup (2002).

Most of the cattle in Uttarakhand are small and non-descript. Crossbred population of cattle accounts for five per cent of the total cattle population. Buffaloes population is largely composed of the grades of Murrah from the neighbouring states. Sheep are mostly concentrated in the upper reaches of the Himalayas. Goats are widely distributed in all districts. Pig population in Uttarakhand is tiny and is only concentrated in the plain and foothill areas of the State around Hardwar, Udham Singh Nagar and Dehradun. Horses, mules and donkeys are the carry pack animals. Over 60 per cent of the poultry birds are desi being reared in the backyard system.

Cattle bullocks are the sole source of draught power in the mountain areas. The plain agriculture in the districts of Dehradun, Udham Singh Nagar and Hardwar, however, is also supported by tractors.

Livestock Genetic Diversity

While almost all the livestock species found in the plains are reared in the mountains, there are four species, *viz.*, yak, mithun, Angora rabbit and ponies, which are reared exclusively in the mountains. Cattle in the mountains are mostly non-descript. However, there are crossbred cattle also with certain proportion of the Jersey and Holstein-friesien blood. They are of different sizes, shapes and colours. Traits change from place to place, particularly in accordance with the altitude of the mountains. Buffaloes are the main dairy animal in Uttarakhand and almost all of them are grades of Murrah from the neighbouring states brought in for milk production.

Asset Value of Livestock

The species wise contribution to the asset value and the unit value are presented in Table 4.5. More than half the asset value of the livestock is on account of the most important milch animals in the State. Adult female buffalo and crossbred clearly indicate the primacy of milch animals in the rural livelihood system. Working males make up the next most important group, accounting for over 30 per cent of the total asset value.

Contribution of Livestock to State Economy

Contribution of livestock sector to the Gross State Domestic Product in 1999-2000 accounted for 7.51 per cent (contribution of agriculture along with livestock being 37.02 per cent) at current prices. Total output value and product wise contribution is presented in Table 4.6. Milk accounted for as much as 77 per cent of the total output value.

Employment in Livestock Sector

Agriculture including animal husbandry continues to be the largest employer in Uttarakhand. Some 70 per cent of the main work force in the state (1.68 mln), was involved with livestock either as owners or as hired labourers (1991 Census). Macro economic studies in other states (1998) report an average of 110 man days of employment per year for household livestock management, per livestock keeping small holder family, in general. This in 2001 added up to some 97.24 million man-days of total employment for the care and management of livestock in Uttarakhand.

Table 4.5: Asset Value of Livestock in Uttarakhand

Sl.No.	*Species*	*Numbers*	*Unit Value (Rs.)*	*Total Value Rs. Mln*	*Per cent*
1	Cattle Indigenous & Crossbred (Male)				
A	Adult Male	734177	3780.00		
B	Young Stock Male (Adult Equivalent)	125076	3780.00		
	Total	859253	3780.00	3247.98	24.97
2	Cattle (Female)				
A	Indigenous Adult	625000	1500.00		
B	Indigenous Young Stock (Adult Equivalent)	175068	1500.00		
	Total	800068	1500.00	1200.00	9.23
3	Cattle (Female)				
A	Adult Crossbred	46301	10000.00		
B	Young Stock Crossbred (Adult Equivalent)	12546	10000.00		
	Total	58847	10000.00	588.47	4.52
4	Buffalo Male				
A	Adult Male	61487	3780.00		
B	Young Stock Male (Adult Equivalent)	87317	3780.00		
	Total	148804	3780.00	562.48	4.32
5	Buffalo Female				
A	Adult Female	655071	8000.00		
	Young Stock Female (Adult Equivalent)	113321	8000.00		
	Total	768392	8000.00	6147.14	47.26

Contd...

Table 4.5–Contd...

Sl.No.	Species	Numbers	Unit Value (Rs.)	Total Value Rs. Mln	Per cent
6	Sheep	310705	800.00	248.56	1.91
7	Goat	1085700	700.00	760.00	5.84
8	Pig	31551	1200.00	37.86	0.29
9	Horses & Mules	23830	7000.00	166.81	1.28
10	Poultry	972269	50.00	48.81	0.38
	Total			13008.11	100

Notes: 1. Stock increment and Capital Gain not included; 2. Young stock of all age groups equated to 0.5 adult units per head for valuation purposes; 3. Crossbred cattle male valued along with indigenous male as their numbers are very small.

Source: Kurup (2002).

Table 4.6: Livestock Sector Output Value: 2000-2001

Sl.No.	Details	Unit Price Rs.	Production	Total Value Rs. mln	% to Total
1	Milk		Metric Ton		
	Cow	8.00 per kg	145235	2323.76	
	Buffalo	11.00 per kg	422602	4648.62	
	Goat	7.00 per kg	1600	112.00	
	Total		**0.569	7084.38	77.15
2.	Meat		Metric Ton		
	Buffalo	30.00 per kg	1860	55.80	
	Goat	100.00 per kg	3050	305.00	
	Sheep	100.00 per kg	1420	142.00	
	Pig	80.00 per kg	980	78.40	
	Total		7310	581.20	6.33
3.	Dung	200.00 per MT	5.94	1188.00	12.94
4.	Wool	25.00 per kg	**233	5.83	0.06

Contd...

Table 4.6–Contd...

Sl.No.	*Details*	*Unit Price Rs.*	*Production*	*Total Value Rs. mln*	*% to Total*
			Numbers		
5.	Hide: Large Ruminants	500 per unit	69350	34.67	
	Skin: Small Ruminants	60 per unit	402131	24.12	
	Total			58.79	0.72
6.	Eggs	1.50 per unit	84.87 Million Nos	127.30	1.39
7.	Work Output	3 per kWh	45.79 kWh	137.38	1.50
	Total			9182.88	100

Source: Kurup (2002).

Notes: Output Value at current price; ** Product quantities are study estimates.

Traditionally, the chores related to care and management of livestock are shared between the men and women in the family. Grazing of animals in commons or forests, movement of animals to markets and handling of animals for AI or health care are the chores for men; while women attend to vital duties like fetching fodder from forests, milking of milch animals, tending fowls, pigs, goats, stall feeding of all animals and cleaning of animal sheds, etc. It was found in a study that about 85 per cent of the total dairy farming work in mountain areas, in terms of the total time devoted, was done by women (Singh and Tulachan 2001).

Draught animal power (DAP) is another area that provides employment, to a certain extent, to large number of families in rural mountain areas. A scenario relating to farm families' engagement in DAP related activities is presented in Table 4.7. While the DAP hiring-out practice is completely absent in the Greater Himalayan zone, it prevails in all other agro-ecological zones. Large landholders, except in the middle mountain villages under traditional farming, do not hire-out bullocks. In the event that all households cannot afford to own pair of bullocks, sharing DAP is clearly a positive indicator, for it tends to keep the population of draught animals in balance, promotes efficient and economic use of the existing population, and also stimulates social cohesion. While hiring-in

DAP saves expenditure on bullock rearing throughout the year, hiring-out, that has been strengthened through the commercialisation of agriculture, has created avenues of employment and income generation for some small and marginal families (Singh 1998).

Table 4.7: Percentage of Different Landholding Groups Participating in Sharing and Hiring of Draught Animal Power (DAP) in Uttarakhand Mountains

Particulars	*Marginal*	*Small*	*Medium*	*Large*	*Overall*
Shivaliks					
Sharing	32.17	26.61	7.14	6.06	25.37
Hiring-in	33.91	35.78	14.29	69.70	37.87
Hiring-out, All Holdings/ Bullock-owning Holdings	12.17/ 36.84	11.93/ 30.95	7.14/ 9.09	0.00	10.29/ 28.00
Middle Himalayas: Traditional					
Sharing	23.08	0.00	6.45	0.00	7.08
Hiring-in	0.00	2.00	22.58	16.67	7.96
Hiring-out, All Holdings/ Bullock-owning Holdings	34.62/ 47.37	24.00/ 24.00	19.35/ 19.35	16.67/ 16.67	24.78/ 24.78
Middle Himalayas: Transformed					
Sharing	24.00	8.70	0.00	0.00	10.67
Hiring-in	52.00	39.13	38.89	11.11	40.00
Hiring-out, All Holdings/ Bullock-owning Holdings	20.00/ 100.00	26.09/ 50.00	11.11/ 16.67	0.00	17.33/ 34.21
Greater Himalayas					
Sharing	11.27	20.00	9.52	0.00	9.17
Hiring-in	50.70	0.00	0.00	0.00	30.00
Hiring-out, All Holdings/ Bullock-owning Holdings	0.00	0.00	0.00	0.00	0.00

Source: Adapted from Singh (1998).

Livestock and Energy

Draught power is the form of animate energy mountain farmers overwhelmingly use in agriculture for operations like ploughing, leveling, puddling, inter-culture, etc. They also provide dung most of which is converted into manure. A proportion of dung is used for producing cooking energy, either in the form of dung cake or as biogas. In mountain areas, use of dung cake is rare, while the same prevails in the Bhabhar and Terai areas of Uttarakhand. Biogas plants are in sufficient numbers in some of the areas. The slurry, as the waste of the biogas plant, is used as manure in the fields.

Availability of energy for cooking food and food security go hand in hand. In this way, livestock have no better renewable alternative. Energy security in mountain areas that can be ensured by animals, therefore, would be instrumental in ensuring food security.

Livestock and Organic Farming

Organic farming, inevitably dependent on livestock for the essential supplies of organic manure, is on the way to a great global revolution. In addition, livestock maintain nutrient transfer and recycling in mountain farming systems that ensures ecological integrity of the whole system (Sing 2002). Development of livestock sector thus would be an imperative for developing organic farming which, in the wake of the declaration of Uttarakhand as an Organic Farming State, is a priority agenda. Organic farming is emerging as one of the vital means to food, health and economic security in a sustainable environment. The fragile mountains of the Himalayas have a bigger say in this respect. Prospects of organic farming in the Uttarakhand mountains are bright because of livestock-farming linkages.

Conclusion

Livestock biodiversity plays a critical role in the framework of food security. Livestock meet direct nutritional requirements through milk, meat and eggs. Their role in agriculture, particularly as draught animals, is instrumental in operating farm activities for food production. The process of ensuring food security thus begins with livestock. Furtner, livestock maintain fertility status of the farms through nutrient transfer and recycling and thus contribute to the essential ecological integrity of a farming system and consequently to agricultural sustainability. Food security, in fact, emanates from this sustainability of agriculture. Contribution of asset and input values of livestock sector to the Uttarakhand economy is impressive. Livestock make important source of earning monitory incomes that takes care of farmers' purchase power, a necessary process of ensuring a family's food security. Prospects of organic farming in mountain areas are bright because of cropping-livestock linkages in this region. Livestock also contribute to equity in the society and this contribution is more pristine than the land and other forms of wealth.

An improvement in livestock sector through effective local level planning with focus on the conservation of biodiversity comprising unique livestock species and genetic resources would lead to an enormous improvement in the socioeconomic condition of local people. Local availability of feeds, livestock genetic resources, health cover, marketing facilities, etc are the most important factors on which local level livestock development planning can be based. Mountain areas have enormous untapped potential for sustainable development of agrarian economies. Natural and livestock resources are such an area that can ameliorate livelihood systems, ensure food security of the masses and revolutionise diverse economies in the region.

References

Gautam, P.L., 2004. Biodiversity for Food Security. *Souvenir: World Food Day*. GB Pant University of Agriculture & Technology, Pantnagar, India, pp. 120.

Kurup, M.P.G., 2002. *Livestock Sector in Uttarakhand & Integrated Livestock Development Plan.* DASP Uttarakhand, Dehradun, India.

Singh, V., 1998. *Draught Animal Power in Mountain Agriculture: A Study of Perspectives and Issues in Central Himalayas, India.* ICIMOD, Kathmandu, Nepal

Singh, V., 2002. Smallhoder Dairy Farming in Uttarakhand, India: Characteristics, Constarints and Development opportunities. In: *Small-holder Dairy in Mixed Farming Systems of the Hindu Kush-Himalayas.* Tulachan, P.M., Jabbar, M.A., and Mohamed Saleem, M.A. ICIMOD, Kathmandu, Nepal

Singh, V. and P.M. Tulachan, 2001. Gender contribution to smallholder dairy production in Uttarakhand. ENVIS Bulletin: *Himalayan Ecology & Development*, 9 (2): 38-39.

Tulachan, P.M., M.A. Saleem, J. Maki-Hokkonen and T. Partap, 2000. *Contribution of Livestock to Mountain Livelihoods: Research and Development Issues.* ICIMOD, SLP, FAO and CIP, Kathmandu, Nepal.

Chapter 5

Traditional Health Management in Indian Himalayas: Ethnomedicine Stays Popular for Livestock Disease Treatment

☆ *Manisha Joshi, Babita Bohra and Vir Singh*

ABSTRACT

Natural ecosystems in the Himalayan mountain areas serve as a rich repository of numerous plants of ethnomedicinal value being used by local livestock-dependent communities in animal health management. This chapter brings forth a comprehensive list of the plants indigenous to the Himalayan mountain areas which carry vital medicinal values and have been exploited by local people for millennia. Folk uses as well as medicinal uses of these plants have been elaborated. In the wake of on-going rapid agricultural transformation and globalization, use of plants of ethno-vet medicinal values and indigenous knowledge woven around the usage of such plants is fast vanishing. These, however, are very much alive in the Indian Himalayan region. Use of these uncultivated/wild plants in the animal production systems by the smallholders in the poor countries of the South, indeed, poses a challenge to the systems operating in the industrialized countries of the North as well as to the rapidly going processes of globalization. Utilisation of these ethnomedicinal plants is cost-effective, affordable and time-tested and needs to be preserved and promoted.

Keywords: *Ethnomedicinal plants, Health management, Himalayas, Indigenous knowledge.*

Introduction

Depending on seasonal differences in temperature and precipitation, pastoralists utilize different pastures in summer and winter, rousing their livestock from mountain areas to valley pastures in winter, or from wet-season to dry season pastures in arid regions. Their pastoral knowledge embraces a whole range of medicinal plants and their usage. Herders often make use of a ample range of Ethno botanical awareness about fodder plants and medical plants for animal use.

The rich and diversified flora of India provides the most valuable storehouse of medicinal plants. The healing properties of herbs have long been known and are documented in ancient manuscripts, such as the Sanskrit *Rig Veda*, *Garuda Purana* and *Agni Purana* (Holland, 1994). These treatises focus on the potential of plants and herbs to cure human ailments and diseases. But the botanical wealth of India also offers the people who tend livestock a rich reservoir which they can tap in their efforts to treat the diseases and ailments of the animals they have for so long depended upon. Given that Indian communities are traditionally rural in nature, a great deal of knowledge in this field has been accumulated over the years. And this indigenous veterinary knowledge is also worth recording. There are no ancient manuscripts comparable to those mentioned above, but scientists are now documenting the various ethnoveterinary practices based on plant drugs (De 1969; Sebastine 1984; Sebastine and Bhandri 1984, Joshi and Singh, 2004). In the mountain areas or in other rural areas of India, pastoral people still keep the traditional veterinary practices based on the remedial power of the plants.

Materials and Methods

This chapter is based on students work during Himalayan excursion. The information has been derived from intensive interaction with farmers in the upper Himalayan areas. The farmers, often the pastoralists, who have been using plants of ethnomedicinal value, are replete with the knowledge about such native plants. Name of the plants and their therapeutic uses as recorded in this chapter are part of their traditional knowledge system.

Results and Discussion

Indian Central Himalayan Region, being the hub of medicinal and aromatic plants, recently attracted the attention of the number of pharmaceuticals hunters ever since the Uttaranchal government declared the region as *Jadi-buti* Pradesh (the State of Medicinal Plants). Numerous plants known for prevention and cure of virtually all human and livestock diseases and health related maladies are thriving in the diverse ecological niches ranging from lower hills to mid-altitude mountains to alpine meadows adjacent to perennial glaciers in the Himalayas. But, only a limited numbers of these plants have been identified and utilized for the cure of a few diseases. A large number of the plants occurring in all the niches are still to be identified, tested against various diseases and utilized on sustained basis.

The Himalayan flora is a unique in the multitude of its values. People inhabiting and managing Himalayan environments are also replete with the knowledge of the utilization of a variety of Plants (Singh and Jardhari, 2001).

Local inhabitants in the Himalayan Mountains would identify a particular endemic plant and elaborate its folk uses and medicinal uses. This ethnic wisdom is unique in itself, but has not been documented properly. An effort has been made to record the names of some of the commonly used plants and their folk uses as well as their medicinal uses for humans and pets (Table 5.1).

It should be mentioned that medicinal plants types change according to habitat type right from lower hills to high Himalayas. This study pertains to the districts of Chamoli which lies towards the interior of the Central Himalayan region. Picture of any other region would be somewhat different.

Regular Practice

Livestock-dependent community farmers have to deal with various health problems occurring amongst their animals, the most

Table 5.1: Ethnomedicinal Valuation of some of the Plants

Botanical Name	*Folk Use*	*Medicinal Use for Humans and Pets*
Berberis asiatica Rox.ex Dc.	Roots are used as a daruharidra for eye disease. An extract prepared by digesting in water sliced pieces of roots, stem and branches is used as Resout and is used in case of opthelmia. In Baluchstan, leaves are administered as cure for Jaundices.	Roots are bitter in taste. They heal ulcers, urethral discharge, skin diseases, ear and eye diseases, useful in ohthalmia, jaundice, disease of mouth and fever, an antidote to snake venom.
Viola serpens Wall.	Fresh leaves and flowers of this plant are collected by local inhabitants and given in the form of tea in asthma and bronchitis.	Used in sprain, roots are given as an emetic. Leaves are used as an emollient and laxative. Flowers have diaphoretic properties.
Angillica glauca	Roots are mixed with condiments for flowering purpose and small pieces are also eaten to cure indigestion	Rootstock is aromatic and used as condiment. It is also used in removing Flatulence and dyspepsia and it has aromatic, stimulant, carminative, diaphoretic and expectorant properties. Rootlets are particularly rich in essential oils and aromatic compounds.
Carum carvi Linn.	Seeds are collected and used as condiments especially for flavoring purposes	Seeds have carminative, anti-emesis, diuretic, expectorant and aromatic properties. The essential oil is extracted from seeds and used in ayurvedic therapy.
Morina longifolia	Pate of roots with water is applied on cuts and boils	Roots are aromatic and locally used as poultice on boils.
Artimisia sacrorum Ladib.	The whole plant is given to horses in head infection. Leaves are used as anthelminthic	The aerial parts yield essential oil which is used in perfumery industries. It is supposed to be anti malarial.

Contd...

Table 5.1–Contd...

Botanical Name	*Folk Use*	*Medicinal Use for Humans and Pets*
Sassurea obvallata Wall.	Flowers are offered in the prayer of lord Shiva. In temples like Rudranath none other this flower is offered to lord shiva. It is also used in Tibetan Medicines in Ladakh.	The roots are applied on cuts and bruises, smell of the flower causes giddiness
Rheum emodi Wall.	The rhizome paste with water is applied on cuts, burns, sprains and in joint pains.	The rhizome is used on cuts and wounds. The root powder is as purgative. The rhizome is commercially collected and sold for purpose of medicines. It is used Tibetan medicines too.
Taxus baccata Linn.	Red aerial part of ripe seed is edible. Leaves are pounded and paste is given orally for treating asthma and other bronchial disorders and indigestion	In recent past a popular drug taxol has been extracted from the leaves and bark of *taxus baccata*. It is being used in curring ovarian cancer in women

Table 6.2: Lists of Plants which are Used by the Farmers

Name of the Plants	*Therapeutic Applications*
Agele Marmelos (bael or bel fruit)	The fruit is roasted and the contents carefully removed. A paste is prepared with water and used as a poultice to treat swollen and painful joints.
Allium cepa (onion)	A mixture of 250 gm onion bulbs and 250 gm solidified jaggery (unrefined brown sugar made palmwine derived from fruit of the toddy palm *Borassus flabelifer* L.) is pounded a dry paste. This is administrated orally each morning for ten consecutive days, to stimulate the virility and reproductive performance of bulls.
Ceiba pentandra (kapok, white silk-cotton tree)	The leaves are pounded together with fermented boiled rice water and extract is administered to cows orally as remedy for reproductive problems. Approximately 500 ml three times a day for three consecutive days.
Caltropis gigantean (gigantic swallow root, *ginat* milk weed, swallow wort)	A handful of leaves are crushed and given orally to cattle to make them more alert and active.
Cissus quadrangularis (quadrangular cissus)	The whole plant is crushed into paste and given orally to newborn calves to facilitate removal of the placenta: after swallowing the paste, the calf will start to vomit.
Datura metel (hindu datura)	A seed paste mixed with water is applied to the patella (knee cap) area in cattle, to relieve pain and swelling. The paste is also used to treat eczema and other skin problems.
Erythrina suberosa	The leaf paste is mixed with water and given orally to cattle as a cure for severe coughing and bronchitis.
Lablab purpureus (hyacinth bean)	The leaf paste is applied to boils and sores to draw out the pus.
Leucas aspera	The leaf paste is applied to wounds to promote healing.

Contd...

Table 6.2–Contd...

Name of the Plants	*Therapeutic Applications*
Luffa acutanagula (chinese okra)	The leaf paste is applied to the neck region of cattle to treat the swelling and sores of yoke gall.
Musa xparadisiaca (french plantain)	The flowers and fruits are crushed and made into a paste with dry ginger (*Zingiber officinale*), pepper (Piper nigrum), black myrobalan (*Terminalia chebula*), nutmeg (*Myristica fragrans*) and karanda (*Carissa carandas*); this is given orally to cattle for all forms of serve diarrohea.
Piper betle (betal pepper)	Ten betal leaves and 20 gm of dry black pepper are made into paste and given orally to cattle as cure for digestive disorders and flatulence. This is repeated two to three times. In addition, in the treatment of cataracts, the person tending the cattle will often chew a mouthful of betel leaves and then spit directly into the animals' eye. This treatment is given in the morning, and repeated on three consecutive days.
Solanum surattense (nightshade)	The leaves of this plant are made into a paste with thuthuvalai (*Solanum trilobatum*) and hot water. The extract is mixed in neem oil (*Azadriachta indica*) and given orally to cattle as remedy for all types of chronic cough.
Tamarindus indica (Indian tamarind)	The leaf paste is applied as a poultice, to reduce pain and swelling in the joints of cattle.
Tribulus terrestris (Ground burnut)	The seed paste is given orally to new born calves to facilitate removal of the placenta.
Zingiber officinale (ginger)	A paste made up which consists of 10gm each of dry ginger, pepper, asafoetida (*ferula asafoetida*), and sweet flag (*Acorus calamus*) in hot water. This extract is administered orally to cattle as cure for gastric problems.

Source: Rajan and Sethuraman (1997).

common of which are fractures and diseases like abdominal discomfort, flatulence and convulsions. Many of these ailments can be treated by the people who tend the animal themselves, because the ingredients or materials required are easily available. In the case of broken bones, for example, the people use bamboo sticks to support the fractured bone. Often they make a fine paste of powdered black gram seeds and egg white, which they place on the site of the fracture or they apply gingerly oil to the spot twice daily.

Some livestock diseases require the assistance of a specialist. Since the government veterinary dispensary is far away, people have to rely on traditional veterinary practices as a first line of traeatment. The traditional knowledge of plant based remedies for the treatment of animal rests with the medicine men, all of whom belong to one family of hereditary indigenous practitioners. Skills and experience are passed on from one generation to the next by word mouth, and are guarded like secrets. The medicine man collects the plants needed for particular veterinary application, either directly from the forest or from the local shops (Rajan and Sethuraman, 1997). Table 2 presents some ethnomedicinal plants and their therapeutic applications.

Conclusion

Ethnomedicinal plants found in diverse Himalayan habitats are impregnated with certain active principles useful in curing virtually all livestock diseases and ailments. These plants should be protected, preserved, grown and promoted in their natural habitats. Efforts must be made to compare the efficacy of medicinal plants with the modern medicines. Active principles associated with Himalayan plants must be isolated and identified. This cost-effective, affordable and easily accessible system of health management must be recognized and promoted in the interest of the marginalized mountain pastoralists.

References

De, J.N., 1969. Further observations on the ethnobotany of Purulia District in West Bengal. *Indian Forester*, 95 (8): 551-559.

Holland, B.K., 1994. Prospecting for drugs in ancient texts. *Nature*, 369-702.

Joshi, M. and V. Singh, 2004. Ethnobotanical valuation of some plants in Himalayan environments. Jothi, P. and K.N. Shukla (ed.), Workshop on *Environmental and Pollution Awareness in Hilly Areas*. GB pant University of Agriculture and Technology Pantanagar.

Rajan, S. and M. Sethuraman, 1997. Indigenous knowledge and development monitor, *IK Monitor Articles* (3-5), ikdm@nuffic.nl, 1-7.

Sebastine M.K., 1984. Plants used as veterinary medicines, galactogogues and fodder in forest areas of Rajesthan. *Journal of Economic and Taxonomic Botany*, 5: 785-788.

Sebastine, M.K. and M.M. Bhandari, 1984. Plants used as veterinary medicine by Bhils. *International Journal of Tropical Agriculture*, 2: 307-310.

Singh V. and V. Jardhari, 2001. Landrace Renaissance in the mountains: Experiences of the Beej Bacho Andolan in Grahwal Himalayan Region, India. *An Exchange of Experiences from South and South East Asia: Proceedings of the International Symposium on Participatory Plant Breeding and Participatory Plant Genetic Resource Enhancement*, 87-96. PRGA, IRDC, DFID, DDS, LI-BIRD, IPGRI and ICARDA, Cali, Columbia.

Chapter 6

Agricultural Transformation and Draught Animal Power System in Indian Central Himalayas

☆ *Vir Singh and Tej Partap*

ABSTRACT

This chapter examines the transformation processes of mountain agriculture in the special context of draught animal power (DAP) system. In large areas of the Central Himalayas predominant traditional subsistence agriculture is in evidence. Majority of mountain population overwhelmingly subsists on and engaged in traditional agriculture. Transformation has reduced the sharing incidence and has encouraged the hiring incidence. Families having inadequate human labour support due to migration of the male persons or otherwise generally do not rear bullocks and depend exclusively on hired plough. Medium and large landholders generally tend to maintain bullocks throughout the year if there is labour surplus at the family or at least during the season. Plough hiring incidence also persists particularly in medium sized group if there is labour shortage. Transformation in mountain agriculture is set to severely affect livestock fodder supplies from cultivated land. Changing trend in the cropping patterns involving high-value commercial crops averts any chance for increasing dependence on the cultivated land for livestock fodder. The land under horticulture-dominated farming system, the major type of transformed agriculture, has almost ceased to provide fodder for livestock. Ongoing trends in the change of cereal cropping pattern are also not conducive to livestock health. Since the dwarf varieties are characterised by narrow straw-grain ratios, their adoption in favourable areas has not assured higher fodder availability. Cattle improvement programme through crossbreeding, popularised as White Revolution, despite huge investment of money, has been met only with little success. This breeding strategy is not compatible with mountain specificities.

Keywords: *Agricultural transformation, Breeds, Bullocks, Draught animal power (DAP), Horticulture, Unsustainability, White Revolution.*

Introduction

Transformation in agriculture has led to alter energy systems in rural areas. Mechanisation has been central to the Green Revolution technology. It gives no value to draught animals for use in agriculture. Although farmers in some highly patronised areas have widely adopted new high yielding varieties (HYVs) irrespective of farm size and tenure, factors such as soil quality, access to irrigation water, and other biophysical and agro-climatic conditions have been formidable barriers to their adoption. Impressive economic contributions of the Green Revolution to large farms in Terai area of Himalayas, to a great extent, are attributable to massive amounts of fossil fuel energy used in farm machines. Farmers of almost all categories in the mountains who have to be content with the HYVs, mono cropping and use of very small amounts of chemicals, have to operate all farming activities using the energy of their animals and their own muscles.

In the following paragraphs, we shall discuss the transformation processes of mountain agriculture in the special context of draught animal power (DAP) system. Discussion on current transformed scenarios would be preceded by that on traditional scenarios and unsustainability indicators and contributing factors. It will be followed by a discussion on limitations households face, changing cropping patterns affecting livestock fodder and the White Revolution technology that has wide-scale implications on DAP.

The Predominant Scenarios

In large areas of the Central Himalayas predominant traditional subsistence agriculture is in evidence. Majority of mountain population overwhelmingly subsists on and engaged in traditional agriculture. High degree of inaccessibility or isolation creates conditions for traditional agricultural management. Higher the degree of inaccessibility the brighter the chances of traditional management. In isolated areas farmers generally tend to be self-sufficient. Natural diversity existing in the traditional systems, as gauged through farming systems, farming situations, cropping systems, plant and animal species, and variability within species, has been and is being utilized by traditional farmers for their sustenance and for developing diverse food production and livelihood systems (Singh 1995).

Production activities in the traditional areas are less diversified. There are two types of farming systems: (*i*) cereal crop-dominated, and (*ii*) livestock-dominated farming system. Cereal crop-dominant system prevails in the Shivalik and foothill zones as well as in the Middle mountains. In general, wheat/paddy-based cropping patterns predominate on irrigated land, maize-based on rain-fed in the hills, millet-based on the upland mid-altitudes and pseudocereal-based on high altitude summer camp lands in the middle mountains.

Croplands in the foothills/valleys are largely irrigated. In the Middle Himalayas it is largely rain-fed and in the Greater Himalayas, it is totally rain-fed. Whereas in the foothill zone both traditional and high-yielding varieties of crops are cultivated simultaneously, in the traditional areas of middle mountains crops mostly include traditional varieties. The herd is cattle dominated. DAP, the dominant component of the animate energy system, is managed through independent, shared, hired-in and hired-out channels of draught animal use system. The linkages with the market system are poor.

The livestock-dominated farming system followed by transhumant pastoralists residing for half a year in the Greater Himalayan zone, includes the herd dominated by ovine species, specially the sheep. Average livestock holding size is very large and high incidence of bullock sharing in the energy use system is observed. Alpine meadows are an important component of the farming system and they are the major source of livestock feeds for about six months in a year. Linkages of this system with the external market are also poor. The livestock based farming system, in fact, is still in a primitive stage of development.

The Unsustainability Indicators

Unsustainability indicators relating to livestock sector in mountain farming system that have emerged over past 50 years are listed in Table 6.1. Decreased livestock holding size (number of livestock per family), reduced proportion of cattle and bullocks in overall herd and draught animals per household, added up with decreased area under forests/pastures which serve as potential source of livestock fodder and consequent reduction in the availability of grazing land due to some obvious reasons, such as construction activities specially in the alpine zone, colonization of exotic plants (*Lantana* spp. and *Parthenium* spp.) at lower altitudes in Shivalik and Foothill zone and conversion of forest areas into cultivated land almost everywhere are some general unsustainability indicators.

Table 6.1: Unsustainability Indicators Relating to Livestock Sector in Mountain Farming Systems (Time frame: 40-50 Years)

Indicators	*Range of Change*
I. LIVESTOCK RESOURCE BASE	
Decreased size of livestock holding	(-) 63-80 per cent
Reduced proportion of cattle in herd	(-) 73-84 per cent
Reduced number of draught animals per farm	(-) 68-40 per cent
Increased proportion of crossbred cattle	(+) 0-6 per cent
Reduced proportion of draught animals in the herd	(-) 13-56 per cent
Decreased area under forests/pastures	(-) 10-40 per cent
Reduced availability of grazing area due to:	
(a) Construction works in the Himalayas	(-) 2-6 per cent
(b) Invasion of exotic plants in lower hills	(-) 20-80 per cent
(c) Conversion into cultivated land	(-) 10-40 per cent

Contd...

Table 6.1–Contd...

Indicators	Range of Change
II. RESOURCE MANAGEMENT	
Increased use of pine needles as bedding material	(+) 50-100 per cent
Use of weeds as bedding material	(+) 0-5 per cent
Emphasis on White Revolution technology of animal husbandry	Medium–High
Emphasis on non-fodder annuals and trees	Low–Medium
Emphasis on HYVs with narrow straw-grain ratios	Medium–High
Increased use of CPRs for non-pastoral activities	10-60 per cent
Replacement of social sanctions of CPR's use by legal measures	Medium–High
Reduced fallow periods for use of PPRs as CPRs	Six months–few days
III. PRODUCTION FLOWS	
Reduced dung production	(-) 63-80 per cent
Reduced manure application in cropland	(-) 60-75 per cent
Reduced production of crop by-products (straws)	(-) 33-70 per cent
Decreased level of concentrate feeds	(-) 73-84 per cent
Reduced milk productivity per head per day of:	
Cow	(-) 33-50 per cent
Buffalo	(-) 50-67 per cent
Higher intensity of plough hiring	(+) 0-40 per cent
Reduced intensity of plough sharing	(-) 7-30 per cent
Decreased availability of suitable woods for tools and implements	Low–High
Increased time spent in fodder collection from CPRs	(+) 200-400 per cent
Increased dependence on human labour for agricultural work	(+) 100-200 per cent
Fodder supplies from:	
Common lands	(-) 50-75 per cent
Private land	(+) 100-150 per cent
Increased bullock work with restless period per day	(+) 67-167 per cent

On the resource management front, there is now increased use in the animal sheds of pine needles with poor biodegradability instead of highly biodegradable leaves of oaks, rhododendron etc. which earlier used to enrich manure quality. Promotion of White Revolution technology involving intensive crossbreeding of local cows with exotic bulls mainly through artificial insemination, which is not conducive to mountain areas' specificities to be discussed soon, should also be regarded as a negative change. In recent years some emphasis by public sector has been laid on monocultures of some food grain crops (such as white-seeded soybean) and trees (such as *Eucalyptus*, *Poplar*, silver oak), which do no provide fodder and instead replace fodder-providing plants. Increased use of CPRs for non-pastoral activities and replacement of social sanctions of their use by legal measures are the negative indicators reducing the options for development of CPRs and consequently of livestock resources. Few years ago a portion of arable land was left fallow for about six months in a year. During fallow period that private land used to function as a CFR on which livestock were freely grazed. Now, due to reduced fallow periods, especially in highly transformed areas, livestock owners have no facility to use PPR as CPR. Emphasis on HYVs generally characterised by narrow straw-grain ratio in place of native plant varieties with reverse characteristic (*i.e.*, wider straw-grain ratio) further creates conditions for reduced supply of fodder from cultivated land.

Now looking at livestock production flow-patterns, one finds considerable reduction in dung production, manure application in cultivated land, straw production, concentrate feeds and milk productivity. Decreased availability of suitable wood for agricultural tools and implements is the other noticeable negative change. Ecological degradation and reduction in the area of commons leads to decreased supply of fodder from these lands, while steep increase in fodder supplies from private land is evident and these are, obviously, negative indicators. Judging from the over-strain on the work bullocks due to continuous and restless work throughout the day with interrupted fodder and water supplies and required care, should also be treated as a negative indicator (increase in working days and thus in total work hours with adequate feeding and care would be a positive indicator). Decreased number in draught animals also leads to increased dependence on human labour for land preparation often resulting into drudgery on women farmers. Higher intensity of plough-hiring and reduced intensity of plough sharing both symptomatic of reduced social cohesion should also be regarded the negative indicators. The negative indicators portraying the dynamism of unsustainability are, of course, not independent. Changes at one level are bound to induce changes at other levels.

Factors Contributing to DAP Decline

Mounting population pressure is one of the principal factors causing environmental degradation in general, and agricultural deterioration in particular. The gravity of the situation can be realised from the very high population density in Central Himalayan region. In 1971, there were 5.22 persons per ha of cultivated land. In 1981 and 1991, this pressure was 5.66 and 7.1 persons per ha of cultivated land, respectively. In the present study, human density on cultivated land ranges from 8.44 in the Shivalik/foothill zone to 14.75 in the Greater Himalayas. In the middle mountains these figures for traditional and transformed areas are 9.87 and 11.62 persons per ha of cultivated land, respectively. In other words, population pressure in our study areas is considerably higher than in the Central Himalayas as a whole.

The average size of per capita cultivated land, an important indicator of the pressure on land, in Central Himalayas in 1991 was 0.14 ha. Per capita cultivated land figures for our study areas are: 0.22 ha in the hills, 0.13 ha in the middle mountains (traditional), 0.09 ha in the middle mountains (transformed) and 0.10 ha in the high Himalayas. In most of the areas in the Himalayas mountains, according to Partap and Watson (1994), per capita cultivated land declined by 30 to 45 per cent between 1960 and 1980 and this occurred despite the extension of agriculture to sub-marginal lands/steep slopes.

There is a myth regarding population pressure in the mountain areas. It is safely stated in official reports that density of population in mountain areas is much lower than in the plain areas. This is simply put on the basis of number of persons per square kilometer of geographical area. The fact that population pressure on cultivated land in the mountains is comparatively higher often remains concealed. Considering from the point of view of mountain specificities, this pressure, virtually, is bound to become intolerably high. Thus the compulsions to expand subsistence farming by clearing forests on commons and reserves, and even on steep slopes, have been very strong in the past.

Human population growth generally favours a corresponding growth in livestock population. However, due to corresponding increase in the number of households, the livestock holding size (number of livestock per household) decreases. This vast population of livestock, in addition to being a major source of family income, nutrition and agricultural power, also contributes to general deterioration of the environment particularly under the circumstances when CPRs are reduced in size, some legal sanctions relating to the use of CPRs prevail, no community system of CPR management exists and animals are deprived of cropland produce (mainly the food grains as concentrate feed). Common land availability per animal head in our study areas ranges from 0.02 ha in Greater Himalayas to 0.34 ha in the Middle Himalayas under traditional type of farming system. This stocking rate is too high to keep pace with the ecological regeneration of the commons. Some pressure on the commons is eased when animals subsist on fodder from agro-forestry systems, reserve forests and alpine meadows.

A general environment has been created for improving milk productivity of cows through the conventional White Revolution technology. But, this technology, discussed in detail under different section, has achieved only a limited success. Change in herd composition the conventional technology has set off is slowly leading to a decline in multipurpose cattle of native breeds and a steep rise in the population of single-purpose and voraciously grazing ovine species and this situation is likely to aggravate the situation of ecological and draught power deficit (Singh 1992).

The Current Scenario

Commercialisation of mountain agriculture represents the efforts of mountain farmers to use scarce land resources more efficiently for gainful employment and increased incomes. The cropping approach is based on cash crop farming and inter-systemic linkages, new forms of diversification (activities), using inputs from science and technology, and building sound upland-lowland linkages (Partap 1995).

Mountain environments provide suitable niches for special activities and products, and harnessing these niches with appropriate location-specific farming activities provides comparative advantage over the plains. The diverse agro-ecological conditions or farming situations prevailing in the mountains form suitable ecological niches for horticulture, floriculture, spice cultivation, and medicinal and aromatic plants. In the process of agricultural transformation, it is the horticulture that has occupied central place. In the lower fertile valleys equipped with irrigation facilities and well linked with market, vegetable-based transformation has taken place. In some areas at high-altitudes, particularly between 1800 m and 2500 m, fruit tree based transformation has captured the core of the transformation process. In selected pockets of high altitude areas, development of apple orchards is a significant change in mountain agriculture. With high moisture regime, this area also provides an appropriate niche for the off-season vegetable farming. Generally apple cultivation and off-season vegetable cultivation go side by side. In lower valleys, change in genetic composition in cereal crops is a common transformation scenario.

Generally, the transformation in any area is not uniform and 100 percent. A number of cropping systems both representing traditional as well as transformed farming are maintained in the same agro-ecological zone and thus transformation actually is the result of diversified farming activities. This mix of cropping practices tilted towards a commercial farming system, in addition to providing on-farm and off-farm employment opportunities and raising economic standard of the households, enhances security of the farmers depending on the farming system. One thing is common to the development of transformed systems mentioned above. They are all energy-intensive. In addition to the use of low to high external inputs, the transformed systems require high amount of animate energy. In mountain areas, owing to specific circumstances, fossil fuel-powered machines have not become the part of transformation process as would be expected in the plain areas. Draught animals and human muscle energy form the only source of energy system to keep the transformation of the farming system going.

Household Limitations

Different production systems within different agro-ecological zones in the mountains tend to adopt a particular transformation process. The local soil conditions, regional climate, cropping preferences, local productivity, profitability considerations, security aspects and sustainability are the main factors driving a particular transformation process. The general process, however, is constrained by household limitations. Only those farmers who have profitable land size, can afford to meet external input and energy requirements and who have some kind of socio-political awareness necessary to get hold on appropriate technology and marketing facilities, generally join the race.

Families with marginal landholding size generally do not maintain draught animals. They would generally depend on hired bullocks for land preparation. Small landholders generally keep bullocks either throughout the year or only during the land preparation period (ploughing season). Hiring-in and hiring-out of draught animals is also a common practice. Sharing of bullocks or plough is practised particularly in untransformed/traditional areas.

Transformation has reduced the sharing incidence and has encouraged the hiring incidence. Families having inadequate human labour support due to migration of the male persons or otherwise generally do not rear bullocks and depend exclusively on hired plough.

Medium and large landholders generally tend to maintain bullocks throughout the year if there is labour surplus at the family or at least during the season. Plough hiring incidence also persists particularly in medium sized group if there is labour shortage. Practice of purchase, sale and exchange of draught animals among all the groups is very liberal.

In recent years DAP has emerged as a source of handsome profitability and employment for some families especially for small and marginal landholding groups. In addition to preparing their own land, these families hire-out their bullocks with or without ploughman on prescribed rates. This practice, further strengthened by the transformation process, has emerged as an income-generating enterprise for a number of households.

Unlike in the traditional areas, labour exchange and DAP-sharing within a community is rare in the transformed areas. And this should be taken up as a negative implication of the transformation process.

Changing Cropping Patterns: Effect on Fodder Production

Transformation in mountain agriculture is set to severely affect livestock fodder supplies from cultivated land. Presently an estimated

35 per cent of livestock fodder supplies are fulfilled from the cultivated land. With continuous pressure on CPRs, dependence on the cultivated land for fodder supplies is likely to increase. But the changing trend in the cropping patterns involving high-value commercial crops averts any chance for increasing dependence on the cultivated land for livestock fodder. The land under horticulture-dominated farming system, the major type of transformed agriculture, has almost ceased to provide fodder for livestock.

Ongoing trends in the change of cereal cropping pattern are also not conducive to livestock health. The total area under agricultural crops has declined over the last decade, *i.e.*, from 1980-81 to 1990-91. A perusal of total area, production and productivity of the food grain crops (Govt. of UP 1993) would reveal that the decline is particularly evident in the case of cereals. Among the cereals, according to mid-1980s estimates (Singh and Naik 1987b), nearly 56 per cent of the total dry fodder derivable from the cropland in Kumaon Himalayas was contributed by two millet crops, the finger millet and the barnyard millet. But these crops together have registered maximum decrease in the area cultivated over the last two decades. Pulses, the second source of good fodder from the cultivable crops, have also shown a decline in the sown area. Soybean is the only pulse/oilseed crop that has maintained an increasing trend in area under cultivation as well as in production. But this introduced crop being given so much emphasis does not provide fodder and usually replaces the millets that also serve as the major fodder crops. Barley, the only cereal that provides dry fodder as well as concentrate, has also registered a considerable decline both in area under cultivation and production.

Increase in the productivity of major crops has been possible thanks to the adoption of HYVs coupled with external agro-inputs. But, since the dwarf HYVs are characterised by narrow straw-grain ratios, their adoption in favourable areas has not assured higher fodder availability. Moreover, fodder to be provided by the HYVs is less preferred compared to that obtainable from the long-stalked native crop varieties. The mountain farmers give great importance to the high fodder yielding native varieties of crops. Poor straw yields inherently associated with the dwarf high grain-yielding varieties has been one of the factors that has had adverse impact in the agricultural transformation process, particularly in the large-scale adoption of new varieties of food crops.

White Revolution Technology: Implications on DAP

Crossbreeding of indigenous cattle by European bulls has, for decades, been the standard institutional approach for improving the production and productivity of cattle in the Himalayan mountains, as also in the whole country, and in fact, all over the Third World. In the Central Himalayan region, crossbreeding programme was initially started in the year 1956 at Vikas Nagar of Dehradun district in Garhwal where Jersey bulls imported from Europe were introduced. This activity was extended to Ranikhet in district Almora in Kumaon region in 1963. In 1969, the crossbreeding programme was taken up at a massive scale by the Indo-German Project (IGADA) at Almora (Agricultural Finance Corp. Ltd. 1987). Despite all efforts and infrastructure developed by all governments for decades, the population of crossbred cattle in Central Himalayas, however, has been almost stagnant.

While the crossbreeding programme has come up with considerable success in the Himalayan Terai, in the hills and mountains the farmers have largely discarded it. In the context of mountain areas, the White Revolution technology can be questioned on the following grounds:

1 The production of crossbreds cannot be sustained without adequate supplies of feed items, like cultivated leguminous fodder and concentrates–the cakes and the food grains. If the already limited area of the cultivated land is spared for livestock production (for raising leguminous fodder) and if a large proportion of food grains goes to the animals (to meet their demand for concentrates), then animals would be in direct competition with human beings.

2 Leguminous fodder crops grow well in irrigated land. In the mountains of Central Himalayas only about 10 per cent of the arable land is under irrigation. If a sizable area of this land is put under fodder cultivation, it might further aggravate the problem of family food supplies in an area that is already food-deficit and imports large quantities of food grains from the plains. In a subsistence-farming situation, setting apart a part of fertile land for the purposes other than food production for family is very unlikely.

3 Native cattle can eat considerably more of the poor quality roughages than the exotic cattle breeds because the gut

content of the local cattle amounts to 33 per cent of the body weight, far larger than that of the exotic cattle in which the gut content amounts to only 20 per cent of the body weight (Orskov 1984). Indigenous breeds convert the roughages supplemented with some quantities of green fodder (grasses and tree leaves in the mountains) more efficiently in comparison to their crossbred counterparts (Nair 1982; Jackson 1985). The native cattle can sustain and produce to some extent by subsisting on poor quality roughages of crop residues on which the exotic ones cannot even survive properly.

4 Crossbreds are more prone to diseases, external parasites' attacks and weather extremes while the farmers in the mountains prefer to keep the type of breeds which are hardy and adapted to local environment and feed resources. The exotics cannot cover long distances, negotiate with rugged, narrow and stony paths and graze on steep mountain slopes, but the local cattle possess all such qualities.

5 Crossbreeding as such is a very expensive technology of livestock development. This not only compels the livestock owners to depend on market system for the purchase of essential feed items–cakes, brans, mineral mixture, vitamins, additives etc.–but also requires a whole veterinary network to take care of the health aspects and manage breeding programmes. The indigenous livestock, on the other hand, consume the feed produced within the system.

6 The local cattle consume biomass of forests and grazing lands and transfer its nutrients to the cultivated land, eat crop residues and recycle the nutrients in the same land on which they grow. The linkages of local cattle with other farming system components are stronger compared to the exotics or crossbreds whose linkages especially with the common property resources are non-existent or are very weak.

7 The crossbreeding policy does not take into account the environment in which the animals have to live and produce (Singh 1994). In developing or less industrialised countries,

the environment cannot usually be controlled. As the animals have many functions here, diversity has survival value. Moreover animals selected on the basis of homogeneity do not fare well here, as is illustrated by the many failures of importation of 'upgraded animals' from industrialised countries to less industrialised countries (Ørskov 1995).

8 Energetic efficiency of crossbred dairy cows under specific mountain conditions is very low (Singh and Sharma 1993).

Female crosses of European (*Bos taurus*) and zebu (*Bos indicus*) breeds of cattle, however, have been well accepted by the farmers in warm climates thanks to the crosses' increased lactation length, higher milk production and earlier age of calving. The crossbred females have also been fairly well received even in the Himalayan Terai owing to their production traits. The White Revolution technology could be successful significantly only in the areas where large chunk of fertile land could be devoted to animal feeding, where good veterinary services could be provided and where there was a mechanical alternative to DAP. Nevertheless, it could be successful at individual family level, not at community level. In the mountain areas' community-based setting, only those programmes are likely to achieve success that are acceptable at the community level, and not those liked by individual families.

The crossbreeding, in essence, is aimed at creating highly specialised animals. Dairy farmers in industrialised countries keep high milk producing cows; beef farmers keep specialised beef animals. Market-oriented approach in these countries has led to homogeneity both of crops and animals, in a new controlled environment (Orskov 1995). But this may not be a case in poor regions. Neither the farmers themselves nor their crops or animals in such regions are specialised. Diversity in the whole system and diversity in every component of the system is the key to their livelihood sustainability. This is especially true of the farmers in the marginalised areas like those of the mountains.

Draught type breeds the mountain farmers maintain are multipurpose ones. They provide manure and power for agriculture and milk for household. Pouring of exotic genes into local cattle, if achieves significant success, would do away the draught power.

There has been little acceptance of crossbred males for draught even in the plain areas because of the general opinion that their working performance is inferior to that of the indigenous cattle since they lack a large distinctive hump and are unable to tolerate high temperatures (Goe 1983; Singh and Naik 1987a). The farmers all over India have been reported to consider crossbred bullocks as uncomparable in stamina, strength and vigour to the local bullocks (Mali *et al.* 1983; Singh *et al.* 1995). The scientists who have worked on draught efficacy drive of the crossbred bullocks have concluded that the crossbred females are good milkers if fed adequately, but when we turn to the field performance of their male counterparts, it is awfully low and these animals are not suitable as work animals (Annaji Rao 1983).

Roy *et al.* (1972) and Anand and Sundaresan (1974) have argued in favour of the crossbred bullocks on the basis of animals' heavier body weight and better field performance between 06.00 and 08.00 hours admitting that attempting to work the crossbreds after 10.00 hours was difficult. Physical efficiency of the bullocks should not be confused with the economic efficiency. A crossbred bullock requires at least 50 per cent more feed than a bullock of indigenous breed; its economic efficiency, therefore, for the same unit of work output turns out to be only two-thirds that of the latter (Rajpurohit 1979). Examination of results on work performance, physiological, haematological, and biochemical responses by Gattewar *et al.* (2000) indicated that indigenous Hariana bullocks exhibited superiority over their crossbred counterparts.

It, however, needs to be emphasised that energy output or draught capacity depends on many other factors, like timely operations, timely availability of inputs, weather conditions etc. and not merely on the type of breed. Despite the fact that local animal breeds have some very useful traits compatible with mountain specificities and, therefore, need to be conserved, the adoption of crossbreeding programme might be conducive to the Terai and foothill areas where fodder availability is plentiful and crossbred males are often used for carting. Due to easy access to all areas in the Himalayan Terai areas, unlike in the mountain areas, all the necessary inputs and facilities could be extended to the crossbred animals.

Acknowledgements

This article is based on an ICIMOD research project on Draught Animal Power in Mountain Agriculture: A Study of Perspectives and Issues in the Central Himalayas, India.

References

Agricultural Finance Corp. Ltd., 1987. *Integrated Cattle Development Project in U.P. Hills*. Bombay : Agricultural Finance Corp. Ltd.

Anand, U. and Sundaresan, D., 1974. "Crossbred Bullocks can Contribute to Agricultural Operations". *Indian Farming*, 24 : 27-29.

Annaji Rao, V. 1983. "Draft Efficacy Drive of the Crossbreds." *Dairy Guide*, 5:17-19.

Gattewar, A.B.: Singh, R.A.; and Yadav, R.S., 2000. "Working Capacity and Behaviour of Crossbred (F1) versus Zebu Cattle Bullocks". *Draught Animal News*, No. 33, 30-32.

Goe, M.R., 1983. "Current Status of Research on Animal Traction". *World Animal Review*, No. 45, 2-17.

Govt. of U.P., 1993. *Sankhyikiya Diary 1983 and 1993*. Lucknow, India : Govt. of U.P.

Jackson, M.G., 1985. "A Strategy for Improving Productivity of Livestock in the Hills of Uttar Pradesh". In Singh, J.S. (ed.) *Environmental Regeneration in Himalaya : Concepts and Strategies*. Nainital, India : CHEA & Gyanodaya Prakashan.

Mali, S.L.; Upase, B.T. and Deshmukh, A.P., 1983. "Draught Animal Power: A National Wealth". *Dairy Guide*, 5:25-27.

Nair, K.N., 1982. "Animals and Tractors in Agrarian Economies : An Analysis of Some Issues Concerning Cultivation Techniques in Agriculture". Paper presented at Seminar on *Maximum Livestock Production from Minimum Land* at Dacca, Bangladesh, 15-17 February, 1981.

Ørskov, B., 1984. "Possible Paths to Real Development". *New Scientist*, Jan 19, 1984, 28-30.

Ørskov, E.R., 1995. "Changing Needs in Cattle Feed". *Down To Earth*, 4(4) : 28-31.

Partap, T., 1995. *High–value Cash Crops in Mountain Farming : Mountain Development Processes and Opportunities.* Mountain Farming Systems Discussion Paper No. 95/1. Kathmandu : ICIMOD.

Partap, T. and Watson, H., 1994. *Sloping Agricultural Land Technology (SALT): A Regenerative Option for Sustainable Mountain Farming.* ICIMOD Occasional Paper No. 23. Kathmandu: ICIMOD.

Rajpurohit, A.R., 1979. "Bovine Feed Availability and Requirement in Karnataka with Reference to Dairy Development Programme". *Ind. J. Agri. Eco.* 30(3), July–Sept. 1975.

Roy, S.R.; Neogi, A.K. and Guha, H., 1972. "Crossbred Bullocks vs Indigenous Bullocks for Draught Purposes under West Bengal Conditions". *Indian Dairyman*, 24 : 66-70.

Singh, V., 1992. *Dynamics of Unsustainability of Mountain Agriculture.* Report of the MFS–ICIMOD Commissioned Study in the Garhwal Himalaya, India. Kathmandu : ICIMOD.

Singh, V., 1994. "Crossbreeding an Utter Failure". *Financial Express*, New Delhi, July 20, 1994.

Singh, V., 1995. "Biodiversity and Farmers in Mountain Agriculture: Experiences from Garhwal Himalaya". Paper presented at the *Beijer Seminar*, Kota Kinabalu, Sabah, Malaysia, May 15-19, 1995.

Singh, V. and Naik, D.G., 1987a. "Animal Draught Power in the Mid-altitude Himalayan Villages". In Pangtey, Y.P.S. and Joshi, S.C.(eds.) *Western Himalaya : Environment, Problems and Development*, Vol.II. Nainital; India : Gyanodaya Prakashan, 755-770.

Singh, V. and Naik, D.G., 1987b. "Fodder Resources of Central Himalaya". In Pangtey, Y.P.S. and Joshi, S.C.(eds.) op.cit. Vol. I, 223-241.

Singh, V. and Sharma, R.J., 1993. "Energetics of Crossbred Dairy Cows in Himalayan Environment". In Singh, V.(ed.) op.cit., 147-164.

Singh, V; Sharma, R.J. and Sharma, M.L., 1995. "Status of Draught Animal Power in Garhwal Mountains". *Adv. Agric. Res. India*, Vol. 3, 173-178.

Chapter 7

Angora Rabbit for Wool Production: Opportunities in Hill Areas

☆ *C.B. Singh*

Angora rabbit farming has emerged as an important income generating enterprise in mountain areas of India during the last two decades. Rabbit occupies a place, which is midway between ruminants and monogastric animals. Like ruminants, it can utilize green fodders and can perform satisfactorily on rations containing as little as 20 per cent of grains. Its efficiency to convert forage into meat and fiber is of special significance as there is abundant vegetation available in the mountains. Rabbits can be reared successfully in a backyard rearing system to raise wool and meat for the family and a small surplus for sale or as a large scale commercial enterprise to supply to large markets.

In India, Angora rabbits are reared in Himachal Pradesh, Uttarakhand, Arunachal Pradesh and Darjeeling hills. There are at present about 25000 Angora rabbits of different breeds being raised by farmers, NGOs and state government agencies, which are producing annually around 150,000 kg. of Angora wool. Mountain economy is mainly based on agriculture and animal husbandry. The agro-climatic conditions and difficult terrains of the mountains are the main factors constraining the production potential of both land and animals. It has been a known fact that, due to shrinking of pasture land and grazing problems, the tradition of sheep rearing is declining while the demand of wool is still there. To meet out the demand of wooly fiber, Angora rabbit farming can play an important role. In addition to this, it can contribute towards improving the quality of apparels as well as upliftment of rural economy.

Specific Advantages of Angora Farming

Besides this, Angora rabbit farming in hilly areas has some specific advantages which are described in the following sub heads.

Suitable Agro-climatic Conditions

Angora rabbit farming is best suited at a temperature range of 10 to 25°C at an altitude ranging from 4000 to 6000 ft. above mean sea level. The agro-climatic and topographic conditions of hilly regions are not only suitable for this enterprise but can generate a sustainable source of income for the rural households.

Simple Food Habits

Rabbits have simple food habits and can be successfully reared on a variety of fodder-based rations. Angora rabbits can efficiently utilize the kitchen waste which can not be of much use to other animals. In addition to this, rabbit is one of tne best feed converters. The feed conversion efficiency of rabbit is next to poultry.

Space Requirements

The space requirement of rabbits is less and an adult rabbit requires approximately three sq. ft. area only. In a 15 × 12 ft. shed about 100 rabbits can be accommodated in multiple tire system for commercial production, which is almost equal to or less than the space required for an adult poultry bird.

Easy Management

The management practices of rabbits are very easy in comparison to other livestock species. A family does not require any out side labour to manage the daily routine practices up to 100 rabbit. Rabbit can be easily managed by each of the family members like women, children, old and disable persons.

Good Substitute of Sheep

Angora rabbit is a good substitute of sheep in at least two ways: Firstly, its production potential is no way less them the sheep, *i.e.* one kg wool/year for sheep and 800 gm for the Angora and, secondly, the quality of Angora fiber is superior than that of the sheep fiber.

Eco-friendly Nature

Most of the animals are considered to be plant destroyer but Angora, when reared in captivity, is eco-friendly. It can be reared on weeds and unwanted vegetation easily. Rabbit farming may be helpful to sustain environment as it can replace hunting of wild animals, like hare, deer and other animal species. Thus rabbit farming offers an opportunity for the environmental conservation as well production and does not contribute to deforestation and injuring the flora and funa of the hilly region.

Droppings are Good Manure

Rabbit droppings contain about four-five times more nitrogen then the dung of cattle and buffalo. Rabbit manure is highly suitable for vegetable and fruit crops.

No Social Constraints

Just like piggery & poultry no social constraints are there for rabbit farming. Moreover, it is being considered as more sophisticated and well accepted by the society of hill people.

Less Physical Labour

Rabbit is very small, simple & harmless animal and can be handled easily. It can be handled by any of the family members. It require less physical labour than required by other livestock species.

Good Quality of Meat

Rabbit meat is of high quality, being rich in protein, low in fat contents and having least moisture percentage. It can be recommended for the heart patients.

Good Feed Converter

Rabbit is a good feed converter. The feed conversion efficiency of rabbit is next to the poultry.

Employment Opportunities

Angora rabbit has a great potential for providing self-employment avenues in hilly areas of the country. It can provide a number of employment opportunities to the unemployed youth, rural women, small and marginal farmers and persons involved in small cottage industry. It can provide about 18 subsidiary industries like:

- ☆ Nursery raising
- ☆ Wool storage
- ☆ Skin processing
- ☆ Cage fabrication
- ☆ Carding & Spinning
- ☆ Skin goods
- ☆ Feed formulation
- ☆ Cleaning & dyeing
- ☆ Meat processing
- ☆ Medicines & vaccines
- ☆ Knitting & finishing
- ☆ Manure & biogas
- ☆ Shearing of wool
- ☆ Sale of wool & woolen garments
- ☆ Research & extension
- ☆ Wool collection
- ☆ Wool farming
- ☆ Consultancy

Constraints

Rabbit farming is climate specific. Only temperate areas of the country are suitable for Angora rabbit farming. Rabbit can be reared between –2 to 32°C temperature with desirable humidity between 60 to 80 per cent . Therefore only hilly areas ranging 4000-6000 ft. above sea level are best suited for Angora farming.

However, states like Himachal, Uttarakhand, J & K, Sikkim and hilly states of north eastern regions of the country have adopted Angora farming at different levels, but could not achieve a significant production level due to several limitations. There are certain constraints which make of this enterprise uneconomical. These are mentioned in the following sub heads.

Lack of High Yielding Germplasm

At present good germplasm is not easily available in the country. Since last import of germplasm in 1997, the stock is continuously being used for breeding. Present stock average has been restricted between 600-800 g wool per animal per year.

Lack of Organized Market

The profitability depends on the factors like location of market, storage, transportation facilities and the grading of farm products. At present, there is no organized market for Angora wool or its products. The middlemen pick up the wool at their discretion and choice of prices. This uncertainty of prices with poor economic return to the farmers affects the overall growth.

Lack of Proper Breeding Strategies

Most of the private farmers do not maintain proper breeding records, which lead to high level of inbreeding. Unplanned breeding strategy, buying/replacement stock from a rabbit farm with improper record management may convert the good germplasm into a mediocre one.

Lack of Adequate Health Care

Some diseases like pasturellosis, hairball, coccidiosis etc. are widely prevalent causing heavy losses. Non availability of vaccines and some important medicines locally also serves as a big cause for the poor disease control in rabbits.

Lack of Adequate Training and Extension Services

Rabbit farming requires a sound practical knowledge of various technical aspects of farm operations. The existing training programmes conducted by various organizations are insufficient to solve the problems of rabbit farmers due to unavailability of trained extension workers in the areas of rabbit production.

Lack of Postharvest Technology

It is difficult to process the Angora fiber on the available machinery in good quality yarn. Non availability of carding, blending and spinning facilities combined with inadequate training to rabbit farmers to process, forces them to sale the raw wool at poor prices.

Risk Coverage

There is no set up for risk coverage through insurance companies for the Rabbit farming, wool and its value-added products. This is also a set back for the rabbit farming.

Strategies for Improvement

The following strategies may be adopted for the improvement of Angora rabbit farming in the country.

Selection of Superior Germplasm

Strategy to improve the breeding potential of rabbit is to select superior individual available at different sources of country. A subsequent selection, maintenance, breeding and performance evaluation are essential to exploit the full potential from the breed. Selection should be based on clip yield, fiber characteristics, reproductive performance and mothering ability of the female. There is a need to replace the old breeding stock by a superior pure breeding stock of known pedigree lines for future genetic improvement.

Dissemination of Superior Germplasm

Selected germplasm based on on-farm production should be disseminated laterally to the farmers through Govt./Non Govt. agencies for upgradation of farmer's flock. Off-farm trails at farmer's field should be conducted to examine the performance, which would help in disseminating the germ plasm in future after refinement.

Feeding Strategies

In livestock enterprise, the feed and feeding management play an important role as it is reflected on production. Similarly, the feeding influences the production potential of Angora rabbits in terms of its wool yield. Rabbits are selective in their feeding habit and avoid dirty feeds. Thus provision of need-based and clean feed is a pre-requisite to avoid wastage of feed and unusual costs. Feed formulation of complete feed pellets utilizing low cost forage/top feeds/concentrates with high energy contents is essential to improve production level from best fed rabbits.

Like other species, the feed requirements in rabbits vary with different physiological stages; the requirement is generally high in young growing and lactating does. Thus it is advocated to feed the rabbits as per their need and to get a higher growth and production.

Since feeding commercial pelleted feed is expensive hence searching of a low cost alternate is always advocated for economic rabbit farming. There are several agro-industrial products, which are available locally in hilly regions, low in cost and can sustain the energy requirements. Thus an approach should be made to identify the locally available, low cost, quality grade agro-industrial by-products which are devoid of toxic principles and growth inhibitors in order to achieve economic rabbit production.

Popularization of Angora Rabbit Farming

Angora rabbit farming is a low cost livestock farming which can be reared in backyard with kitchen waste. The breed is highly prolific and due to shorter generation interval a large no. of offspring's can be obtained by a shorter spell of time. The rabbit manure is very often utilized efficiently for fruit and vegetable crops. Therefore, there is a strong need to popularize the species among the farmers for up liftment of socio-economic condition of hilly people through Angora rabbit farming.

Exploration of Marketing Opportunities

Presently, there is no organized market of Angora wool and of its products in the country and due to this there is a discouragement to the rabbit farmers. Therefore, best suited strategies to overcome such problems of farmers should be prepared so that discouraged farmers get some boost and reinitiate Angora farming.

Loans/Subsidies

The majority of house holds in rural areas of hilly states are categorized under small and marginal land holding size. The farmers, by and large poor, have limited economic options and have no money to initiate any enterprise on their own cost. Thus, the initiatives have to be taken to provide loans/subsides from Govt./NGOs/Banks/Private financing agencies to encourage farmers to come forward for rabbit farming. Such initial boost would definitely help farmers to show their interest on this profitable enterprise.

Hygiene Management and Disease Control Measures

Health management and control measures are the two basic pre-requisites to achieve higher production. Rabbits should be offered routine health care to reduce mortality in adults, young as well as newly born kits.

References

Bhasin, V. and Singh, R.N. (1998). Strategic approach to maximize rabbit production under Indian conditions. *Golden Jublee Seminar on Sheep, Goat and Rabbit Production*. Jaipur, 24-26 April, 1998. Seminar Compendium, 59-64.

Risam, K.S. and Das, G.K. (2004). *Strategies for improving Angora rabbit production in the country. National Seminar on Angora Rabbit Wool and Cashmere Production and Utilization*, Manali, Himanchal Pradesh, 25-26 Sept. 2004, Seminar compendium, 52-54.

Sawal, R.K. (2004). Nutrition and feeding management of rabbits for fiber production. *National Seminar on Opportunities and Challenges in Nutrition and Feeding Management of Sheep, Goat and Rabbit for Sustainable Production*, Avikanagar 10-12 Feb. 2004. Seminar compendium, PP 148-162.

Sharma, R.J. and Sharma Shingini (2006). Opportunities of self employment through Angora rabbit farming in the hills of Uttaranchal. *National Seminar on Angora Wool Development Programme in Hill Areas of Uttaranchal: scope and opportunities.* Hill Campus, Ranichauri, Uttaranchal, 5-6 Sept. 2006. Seminar Compendium, 55-56,

Singh C.B. and Dev Chandra (1998) Problems of Angora Rabbit Farming in Garhwal Himalayas. *Mountain Ecosystem: A Scenario of Unsustainability*. Indus Publishing Company, Delhi, 160-161.

Singh C.B.; Jilani M.H. and Tripathi R.K. (2004). Constraints analysis of Angora rabbit farmers in Garhwal Himalayas. *Pant J.Res.*-2(2): 76-78.

Singh C.B. and Jilani M.H. (2005) Mortality pattern in Angora Rabbits under temperate conditions of Garhwal Himalaya. *Pant J. Res.* 3(2): 74-75.

Chapter 8

Techniques for Better Utilisation of Crop Residues for Ruminant Diet

☆ *R.S. Jaiswal, Chetna Bhatt and Richa Pathak*

Introduction

Crop residues like Wheat bhusa, Rice straw, Jowar, Bajra and Maize stover are the staple livestock feeds in India and other countries of South-East Asia. These straws are called low quality fodder because of their low nitrogen and mineral contents and the presence of large quantities of lignin & silica and they are poor in digestibility and as such cannot meet even the maintenance requirement of animals. Because of increasing livestock population density and decreasing grazing areas, the straw feeding is becoming more and more important in ruminant diet. About 50-60 per cent dry matter intake in large ruminants in our country is through the crop residues. Ranjhan (1997) indicated that low quality of roughage has both physical and chemical constraints to obtain optimum animal production. The ruminants on straw diet alone lose body weight (Rakib *et al.*, 1970). Thus improved utilisation of these crop residues is very essential. A number of ways have been tried during the last

50 years to improve the feeding value of these crop residues. These methods range from simple supplementation with deficit nutrient to treating them with alkali or combination of physico-chemical or physico-chemical-biological methods.

Treatment of straw with urea as a source of ammonia has now been found to be effective in improving the nitrogen content, intake and digestibility of the treated straws. The urea treatment works out to be cheaper because of the lower cost, easy availability, easy handling, being free from pollution hazards etc.

Treatment of Straws (Loose Form) with Urea as a Source of Ammonia

In this process certain quantity of urea solution is sprinkled on straw so as to result in a moisture content of 50 per cent in case of chaffed rice and wheat straw and 40 per cent in case of unchaffed rice straw. The moisture content of dry wheat and rice straw mostly varies from 15 to 20 per cent. Determine the volume of urea solution required for every kg. of straw. It can be worked out by subtracting the moisture content of straw from 50 per cent in case of chaffed rice and wheat straw and 40 per cent in case of unchaffed rice straw and dividing it by 50 and 60 respectively. For example, the volume of urea solution required per kg. of wheat straw having 18 per cent moisture or unchaffed rice straw with 20 per cent moisture would be

$$\frac{50-18}{50} = 0.64 \text{ and } \frac{40-20}{60} = 0.33 \text{ litre, respectively.}$$

Now the amount of urea required for these straws could be worked out by using the following formula:

$$\text{Urea solution (kg urea/100 litre solution)} = \frac{\text{DM\%} \times 5\text{kg}}{\text{Volume of urea solution required/100 kg straw}}$$

For wheat straw and chaffed rice straw and unchaffed rice and wheat straw about 6.0 and 12.0 kg urea/100 liter of water.

For treating straws, a known quantity of straw is spread thinly on the ground either in the open or in silo or in shed. Suppose an average head load weigh 20 kg., then sprinkle uniformly 12.8 litres of 6.5 per cent urea solution chaffed wheat and rice straw and 6.6

litres of 12.1 per cent urea solution in case of unchaffed straws. Repeat the process layer after layer and also trample the treated straw to exclude as much air from it as possible. The treated straw could be stacked and stored in any of the traditional ways, *viz.* burzi, banga, dhar or simple stack in a part of a room or hut. The only care should be taken is to protect it from rain water which could be achieved by either covering the stack with Kans grass or by plastering it with a mixture of mud and cow dungs. The treated straw is ready for feeding after 3-4 weeks. In case of unchaffed rice straw, the stack is made in the open and the straw is removed from the top, while in case of chaffed/broken straw stack is made inside a shed; it is removed from one side without disturbing the rest of it. The treated straw can be used for about one year provided the stack is protected from rain water.

Treatment of Baled Wheat or Paddy Straws with Urea as a Source of Ammonia

Most of the work related to urea (ammonia) treatment of straw has been conducted on chaffed straws by packing the material in earthen pits or bamboo baskets, stacking in the open or storage in hut type structures. Wheat/Paddy straw could be collected and baled from behind the combine harvester with the help of a field bales, which produced rectangular bales of 36 × 40 × 110 cm. size with a bulk density of straw to 120-150 kg/m^2. For the treatment, 216 straw bales (about 4 tonnes) were stacked on the plant in 6 layers comprising of 36 bales in each layer. The total dry weight of stock was calculated and the amount of urea solution to be added into the stack to raise moisture content upto 40 per cent was determined as 2088.5 kg. The amount of urea was then calculated as 5 per cent (wt/wt.). The urea solution was then prepared with an average concentration of 83.3 g urea per litre of water for 2088 litres of water 174.0 kg of urea was added. The urea solution was prepared in 1000 litres capacity water tank and urea solution was dripped into the straw stacks placed on the treatment plant.

Effect of Urea and Protein Supplementation

Thirty crossbred heifers between 7-18 months of age and 62 to 16.4 kg weight were distributed randomly into six groups of 5 animals each and fed diets based on, *viz.*, T_1–Ammonia treated rice straw (ATS), T_2–ATS plus fish meal, T_3–ATS plus cotton seed cake, T_4–ATS plus dried leucaena leaf meal, T_5–untrcated rice straw plus concentrate mixture to fulfil energy and protein requirements and T_6–untreated straw with urea added at the time of feeding to raise the nitrogen content in the ammonia treated straw. In all groups the straw was fed *ad lib* and protein supplement was given in amounts to supply 30 per cent crude protein of the total crude protein requirement for normal growth. Each heifer was given 2.0 kg green berseem, 30 g mineral mixture and 30 g salt/day. The ammonia treated rice straw was prepared by spraying 65 litres of a 6 per cent urea solution on 100 kg of chaffed straw and stacking it for 4 weeks in shed.

The dry matter (DM) and digestible dry matter (DDM) intake of ATS diets (except fishmeal supplemented) were significantly higher than those of untreated straw plus urea supplemented (Table 8.1). The intake of digestible dry matter of a diet consisting of untreated straw plus concentrate was slightly lower compared to the urea treated straw diet. The dry matter digestibility of the complete diets was not significantly different between treatments. The dry matter digestibility of the straw component of the untreated straw and concentrate diet was significantly lower than that of the ammonia treated or urea supplemented straw component of the other diets. Acid detergent fiber (ADF) digestibility of ammonia treated straw diets was significantly higher than both the untreated straw diets supplemented with urea or concentrate. The effect of ammonia on ADF digestibility is apparently due to solubilization of the hemicullulose (Klopfenstein 1976). Addition of concentrate to untreated straw diets significantly depressed ADF digestibility probably due to the effect of easily fermentable starch and fibre digestion (Jackson 1978). The highest livestock gain was obtained by feeding untreated straw plus concentrate (1.8 kg/d). Among the treatments receiving urea (either at the time of feeding or to generate ammonia) ammonia treatment plus protein supplements gave higher ($P < 0.05$) live weight gains than untreated straw plus urea.

Problems of Technology Adoption

The size of stack is normally determined by the number of animals and duration for which the crop residues are to be used on should normally not make a stack of less than a ton because the smaller stacks have proportionately large surface area which is responsible for the escape of ammonia from the stack. Feeding treated straw for a few days does not bring out the impact of treatment. Thus the minimum size of stack should be such that all the animals can be

Table 8.1: Effect of Urea and Protein Supplements Added to Untreated and Ammonia (Urea) Treated Rice Straw an Digestibility, Intake and Growth of Crossbred Heifers

Parameters	*Untreated Straw*		*Ammonia (Urea) treated straw*				
	Urea	*Concentrate*	*Nil*	*Cotton Seed Cake*	*Leucaena Leaf Meal*	*Fish meal*	*Sem±*
Feed intake (kg)/100 kg body weight/day							
Total dry matter	2.83^{bc}	3.2^{bc}	3.96^{a}	3.92^{a}	3.80^{a}	3.62^{ab}	± 0.20
Straw dry matter	2.83	1.90	3.96^{a}	3.69^{a}	3.50^{a}	3.50^{a}	± 0.20
Digestible DM intake	1.26^{b}	1.60^{ab}	1.70^{a}	1.80	1.80^{a}	1.66^{a}	± 0.11
Digestibility (per cent) Total dry matter	46.7^{a}	31.9^{b}	43.2^{a}	43.7^{a}	45.8^{a}	44.1^{a}	± 2.20
Straw dry matter	44.7	49.4	43.2	45.4	46.9	45.8	± 1.40
ADF	34.7^{b}	20.0^{c}	41.9^{a}	41.0^{a}	43.5^{a}	44.4^{a}	± 1.80
CP	60.8^{a}	53.4^{a}	26.9^{c}	40.0^{b}	42.5^{b}	27.9^{c}	± 8.10
NDF	49.7^{c}	50.4^{bc}	53.3^{abc}	53.3^{abc}	54.6^{ab}	57.7^{a}	± 1.60
Live weight gain g/d	111.00^{d}	473.0^{a}	246.0^{cd}	427.0^{ab}	290.0^{bc}	327.0^{abc}	± 56.0

Source: Jaiswal *et al.*, 1988.

fed from it for at least 5-6 months. About six ton stacks have been found to be convenient by some of the dairy farmers. A circular shaped stack due to its smaller surface area should be better than a rectangular stack. The dome shape at the top is essential to prevent rain water from entering into the stack.

If properly compacted and protected from rain water, the stack once opened could be used for 4-5 months without much change or spoilage. Care, however, is taken not to unnecessarily disturb the stack while taking out the treated straw.

Lack of availability of human labour for urea treatment just after the harvest of paddy crop is another constraint because this is the time when farmers are busy in the preparation of their field for sowing the Rabi season crops.

The fungal growth in the stack of urea treated paddy straw was observed by the farmers in few cases. It was the apprehension of the farmers that if such straw is fed to the animals it might cause some disease and prove harmful to the animals. Therefore, proper identification of fungus, its cause and effect need to be thoroughly investigated.

Sticky durg from animals given urea treated straw is produced because of which farmers require more time to clean the floor. The pungent smell of ammonia puts off some farmers to use treated straw for animal feeding.

References

Jackson, M.G., 1978. *Treating Straw for Animal Feeding*. FAO. Animal production and Health paper No. 10, FAO: Rome.

Jaiswal, R.S., Verma, M.L. and Agrawal, I.S., 1988. Effect of feeding various protein supplements added to urea treated rice straw an nutrient intake, digestibility and growth in crossbred heifers. *Indian J. Anim. Nutri.*, 5(3): 188–194.

Klaptenstein, T., 1976. Chemical treatment of straw residue. In: *Symposium on Crop Residues in Beef Cattle Production System.* American Society of Animal Texas.

Ranjhan, S.K., 1997. Feeding strategies for supporting high livestock and poultry product. In: *National Symposium on Feeding Strategies for Ecofriendly Animal Production in India.* 14-15 Feb.

Chapter 9

Livestock-Mediated Agroecosystem Services in the Himalayan Mountains

☆ *Vir Singh*

The concept of "ecosystem services" or "environmental services" is becoming increasingly popular since the last decade of the last Century. The term actually was coined by the economists and now it is being heard all over and gets frequently encountered in documents generated by several national and international development agencies. Vocabularies of scientists, academicians, NGOs and development departments of public sectors, international organizations are now replete with the phrase "ecosystem services" or "environmental services".

The emerging concept of ecosystem services, in actual sense, provides critical means of taking privatization to a new level–a means of privatizing many things that have as yet been unavailable for privatization: air, water, and all sorts of other ecological processes (GRAIN 2005). It is nevertheless necessary to go deeper into the many aspects of ecosystem services so that we could help ourselves develop a sound perspective compatible with our specific geo-ecological and socio-cultural framework.

In the deliberations on ecosystem services role and functions of livestock are invariably missed. Animals are part of an integrated system with crops or trees and their role and function is often much wider than simply as providers of milk and meat, as in developed countries (Ørskov 2005). Moreover, in mountain areas, sustaining agriculture without livestock is virtually unimaginable (Singh 1998). This chapter attempts to focus on the ecosystem services as they are mediated by livestock in mountain agro-ecosystems.

While an agro-ecosystem is the term generally used for the cultivated area, in the specific mountain context it encompasses cropland, uncultivated forest/rangeland, livestock and households. Management of an ecosystem rests on the will and command of human beings particularly depending on that ecosystem for their livelihood. Livestock management in an ecosystem phenomenally influences the flow of services emanating from an agro-ecosystem.

Soil-Plant-Livestock Interactions

Domesticated animals are part and parcel of mountain agriculture. Given the specific geo-ecological features of the Himalayan mountains, it can be easily inferred that farming in this region cannot sustain without the critical functions livestock perform. Livestock are regarded merely to be a source of products of high economic value as well as of supplementing family income. That they, in addition, also play crucial role in generating the natural ecosystem services in the mountains is seldom appreciated and recognized.

Livestock-forestry-farming linkages are one of the essential features of mountain agriculture. These linkages in the mountain agro-ecosystems are indistinguishable with the commonly talked about "soil-plant-animal relationships" in an ecosystem. This relationship–by means of appropriate flows of nutrients in the ecosystem–generates processes and sustains functions that are vital for soil fertility management and overall natural health maintenance of the ecosystem.

Considerable transfer of nutrients from ecologically more stable forest ecosystem to the more fragile croplands is mediated by livestock through this relationship. Forests, especially the natural ones, are the rich repositories of the nutrients which become a subsidy for cultivated land. These nutrients are transferred to the cultivated land via livestock. The nutrient transfer takes place in two ways. Forest biomass–tree leaves and ground flora–are fed to the livestock. The biomass is also used as bedding material in livestock sheds. Both dung and bedding material are converted into manure which is transferred to the cultivated land. The livestock also recycle the nutrients in the cultivated land. Crop residues are fed to the animals and thus the nutrients in them are recycled into the cropland. The biomass transfer and nutrient cycling mediated by livestock infuses vitality in the production system and livestock themselves fulfill their requirements for maintenance and production (milk, draught power, etc.). This dynamic relationship amongst forest ecosystems, livestock and croplands is vital for the very sustainability of mountain agriculture (Singh 1998).

This service rendered by livestock is crucial for designing mountain ecosystems for continually producing and flowing foods and other essential items of human livelihood systems. Biodiversity in a forest/rangeland ecosystem has enormous bearing on the performance of both livestock and food crops.

Livestock-induced Sustainability

Sustainability, in essence, is the function of a complex of ecosystem services through which a system or a sub system performs at a well defined level without deteriorating itself. Optimisation of the operating factors embroiling into ecosystem services is necessary for managing the state of and conditions for sustainability. Although a number of factors are required to manage sustainability, there are three basic principles to operationalise sustainability in agricultural systems as elaborated by Werf (1993), *viz.* (*i*) biodiversity and complexity; (*ii*) living soil; and (*iii*) cyclic flow patterns. Livestock render services which are vital for the sustainable performance of an agro-ecosystem.

Livestock enhance biodiversity through draught power input which is necessary to cultivate and utilize biodiversity in relay fashion. Complexity is created by using a farm element to its optimum effect through using it as fully and intensive as possible by way of internal cycling and recycling (Werf 1993). Mountain farmers have evolved a system in which they would use a farm produce or element at several levels and in several functions, making waste/intermediate product or function the input for another and ultimately cycling and recycling the end product into the farming system. In this way, elements in large number are placed into a web of symbiotic and dynamic interrelationships. Larger the number of functions fulfilled by a single farm elements, the higher the degree of complexity and sustainability (Singh 2002). The combination of diversity and complexity–high number of species and their genotypes functioning in a web of symbiotic interrelations–results in a stable farming (Werf 1993).

Rumen as an Ecosystem

Rumen–one of the four parts of livestock stomach–is a unique ecosystem in itself. It is a sort of the ecosystem in which microbial life (protozoa, bacteria, fungi) is dominated. It is, therefore, often termed as rumen microbial ecosystem. The rumen ecosystem contains bacteria (10^{10} cells/ml, representing more than 50 genera), ciliate protozoa (10^4–10^6/ml, from 25 genera), anaerobic fungi (10^3–10^5 zoospores/ml, representing 5 genera) and bacteriophages (10^8–10^9/ml) (Hobson 1988). For extracting energy from the lignocellulosic feeds sources available in nature, the ruminant depends upon the microbial ecosystem it harbours in its body. The microbial rumen ecosystem bioconverts recalcitrant lignocellulosic feeds into volatile fatty acids, which serve as energy source for the livestock (Kamra 2004).

Apart from giving a refuge to numerous microorganisms in their rumen ecosystem livestock extend very crucial service of converting lignocellulosic feed sources into useful products that are of great value in human systems.

Biodegradation rates in nature, particularly in the temperate environments as of the Himalayan mountains, are very slow. Rumen ecosystem accelerates the rates of biodegradation. The rumen ecosystem is capable of converting nitrogen of non-food sources such as urea, excreta, nitrates etc. into high quality proteins.

Livestock and Carbon Sequestration

The fields in which the greatest practical implementation of the ecosystem services are the sequestration of atmospheric carbon, the capture and storage of water, and biodiversity, and landscape conservation (primarily for tourism) (GRAIN 2005).

Table 9.1: Livestock-Mediated Ecosystem Services in Managing Agricultural Sustainability in the Mountains

Operational Condition	*Livestock Services*
Diversity-Complexity	☆ Draught animal power enabling various agricultural operations is necessary for cultivation of agrobiodiversity.
	☆ Manure prepared by livestock is a vital input for improving soil texture and managing soil fertility as an environment for cultivating biodiversity and improving resilience.
	☆ Livestock themselves are component of biodiversity.
	☆ Livestock give way to full and intensive use of farm elements and thus contribute to increase functional diversity and enhance ecosystem complexity.
	☆ They help in designing multiple cropping and versatile crop rotations.
	☆ Rumen of ruminant animals is a wonderful ecosystem harbouring variety of microorganisms (bacteria, protozoa, fungi) and functioning uniquely in nature.
Soil Fertility Management	☆ Dung/manure conserves moisture in the soil, improves soil texture and builds up a culture in the soil that harbours numerous life forms in the soil ecosystem.
	☆ Soil-plant-livestock relationships are vital for maintaining.
	☆ *In-situ* manuring during grazing, ploughing and tethering substantially improves soil fertility status.
	☆ Soil fertilization due to urine passed by working animals.
Cyclic Flow Patterns	☆ Livestock manage fragility of croplands through fetching nutrients from ecologically more stable ecosystems to more fragile lands.
	☆ Nutrients of cropland are recycled into the soil through manure.
	☆ Livestock maintain the essential ecological integrity of an agro-ecosystem.

Carbon trading is currently the most common and well known form of the developments taking place to encounter environmental problems as well as to spin dollars. For any company in the United States the cost of reducing emissions at the source may be up to US$ 150 per tonne of carbon; a company in Europe may need to invest up to US$ 200 for the same reduction. Alternatively, the companies can pay other companies and groups , mostly from non-industrialized countries, to reduce emissions or to absorb CO_2 from the atmosphere, and account that as their own reductions. The business of buying and selling these carbon credits has become so big, that 'carbon bonds' are sold in the stock market (GRAIN 2005).

Ecosystem services relating to CO_2 sequestration value preventing deforestation instead of plantations (Singh 2005). Whereas open oceans and natural forests are the major sinks of CO_2, agro-ecosystems too play a significant role in the absorption of CO_2. Agroforestry systems, orchards, silvi-pastoral systems and all other types of food production systems in order to keep thriving need inputs generated by livestock, especially dung/manure and energy (draught power) as well as the invisible services rendered by them. Livestock thus help trapping enormous amount of CO_2 in trees, shrubs, crop plants and grasses. Forest-Crop-livestock integrated systems as in mountain areas are certainly more efficient in absorbing CO_2 than the simple crop monocultures in the plains, particularly in the Green Revolution agriculture.

CO_2 in the atmosphere has increased dramatically in the last 40 years from 316 ppm to 367 ppm, a jump of 17 per cent (Share International 2000). The problem cannot be effectively addressed without making agroecosystems as the main target. Carbon sequestration in agro-ecosystems is a dynamic process. Livestock do extend their services to accelerate the process of CO_2 absorption into plant biomass and soil carbon pools. Soil carbon pool is twice as big as biomass carbon pool. Agroecosystems can contribute to manage the soil carbon pool more effectively.

Future of Livestock Enabled Services

The on-going rapid processes of globalization are all set to extend market system to all spheres of life and transform everything of direct or indirect human use into a marketable commodity. The livestock-based lifestyles will also not be spared. As Ørskov (2005) judges it, there are two systems relating to livestock production systems, *viz.*, (*i*) social value-oriented systems as prevailing in the developing countries particularly in South Asia, and (*ii*) market-oriented systems as are seen in the industrialized countries.

In the former system risk minimization, family support and stability and sustainability are the major goals of rearing the livestock, whereas maximum productivity and profit are the main motives in the latter. As the social values get eroded (or become a matter of the market system), scientists' role, targets, approach and emphasis would also be changed (Table 9.2).

Ørskov's (2005) critical observation can also be judged from ecosystem services rendered by the two contrasting livestock production systems. As the social cultural values vanish, our relationship with nature also gets adversely affected. Ethical and aesthetic values inherently associated with the social value-oriented systems value ecosystems and resource conservation. They owe sound ecological basis, are frugal, sustainable and futuristic. Hence they ensure ecological balance which is an imperative for the flow of ecosystem services on sustained basis. As we dissociate ourselves from the value-based system to fall prey to market-oriented systems, what we are sure to loose is the benevolence of natural ecosystems that they could bestow on us through their aesthetic environmental services.

Ethical Basis of the Valuation of Ecosystem Services

Contribution of livestock-mediated services to the Himalayan communities and–through ecosystem linkages–to the foreland communities in the plains is not an easy job to evaluate in monitory terms. We can ponder over an attempt made for the ecosystemic evaluation of the millennium in 2002. The main motive was to evaluate the costs of compensation for rural communities (http://www.prisma.org.sv/pubs/CES_RC_Es.pdf) quoted in *Seedling* by GRAIN (2005). These are categorized into supply, regulation or control, cultural and back-up, or supporting activities. On the basis of these, an attempt has been made to list the specific contributions of livestock towards ecosystem services in the context of the Himalayan mountains (Table 9.3). A try to evaluate the livestock-mediated services on the basis of these attributes can be given as in the Table.

Table 9.2: Livestock in the Market-Oriented and Social Value-Oriented Systems

Particulars	*Market-oriented Systems*	*Social Value-oriented Systems*
Overall goals for livestock	☆ Profit maximization	☆ Risk minimization
	☆ Cash generation	☆ Family support
	☆ Productivity	☆ Stability and sustainability
Scientists' role	☆ Design of systems	☆ Management of ecosystems
Intermediate targets	☆ Genetic homogeneity	☆ Biological diversity
	☆ Increased production potential	☆ Improved maintenance potential
	☆ Single purpose animals	☆ Multi-purpose animals
	☆ Nutrient mobilisation	☆ Nutrient storage
Philosophical approach	☆ Specialised	☆ Holistic
	☆ Fragmented	
Scientific approach	☆ Single discipline	☆ Multi-and-trans-disciplinary
Statistical emphasis	☆ Mean	☆ Variance
	☆ Main effects	☆ Interactions
Natural consequences	☆ Ecological deficit	☆ Ecological balance
	☆ High risks	☆ Resilience
	☆ Unsustainability	☆ Sustainability

Source: Adapted from Ørskov (2005).

Table 9.3: Livestock Contributions to Ecosystem Services in the Specific Contexts of the Himalayan Mountains

Ecosystem Services in Relation to:	*Livestock Contributions in the Mountain Context to:*
Supply Goods produced or provided by ecosystems such as foods, water, fuel, fibre, biodiversity or natural medicines	☆ Biodiversity (species and breeds) ☆ Manure ☆ Draught power (ploughing, leveling, puddling, inter-cultural operations. ☆ Carry packing/transfer of produce and useful luggage ☆ Threshing ☆ Milk, milk products, wool, meat, eggs, hides, furs, etc. ☆ Plants of ethnovet value
Regulation, or Control Services obtained by regulating or controlling ecosystem processes, such as the quality of the air, the climate, water (distribution and quality), erosion, the causes of illness, the manipulation of biological processes, risk reduction and so on	☆ Managing agro-forestry systems having a number of deep-rooted trees, shrubs and perennial grasses ☆ Maintaining forests/rangelands as the core components of mountain farming systems with primary objective to conserve fodder species in association with associated species ☆ Increasing resilience of the agro-ecosystems ☆ Ecological integrity of the agroecosystems ☆ Improving/maintaining quality of environment and its products as would stem from the above points ☆ Preventing/controlling diseases by means of ethnovet practices ☆ Minimising or averting risks to the farming communities

Contd...

Table 9.3–Contd...

Ecosystem Services in Relation to:	*Livestock Contributions in the Mountain Context to:*
Cultural Aspects The non-material benefits that enrich the quality of life such as cultural diversity, religious or spiritual values, knowledge (traditional or formal), inspiration, aesthetic values, social relations, a feeling of place, the values of a group's cultural patrimony, recreation and ecotourism	☆ Sanctity associated with cattle, the *godhan* ☆ Aesthetic values associated with bullock fairs, thanksgiving to bullock festival (*e.g.*, *Baldraj* in Garhwal Himalayan mountains), *Goibardhan* festival, worship of cow and other rituals ☆ Cultural attributes ☆ Social cohesion in the community based systems of the livestock-dependent societies ☆ Native breeds with unique and superb traits developed in the specific mountain environments and environmental attributes in the maintenance and development of the native breeds ☆ Unique diversity in the cultures of livestock-dependent communities ☆ Recreation values emanating from livestock production systems (*e.g.*, natural bounties such as forest-capped landscapes, alpine meadows), festivals and rituals ☆ Ecotourism in mountain ecosystems–as parts of livestock production systems–managed by livestock-based native communities
Back-up, or Supporting Activities The services required to produce the other services, including primary production, the formation and/or fixture of soil, oxygen production, pollination, habitat creation, nutrient recycling and so on	☆ Services to produce cereals, pulses, oilseeds, medicinal and aromatic plants, grasses, trees, fruits, vegetables, vegetable seed production, etc. ☆ Soil enrichment by means of leguminous plants' cultivation using livestock energy and through livestock-mediated nutrient cycles ☆ Pollination (by birds) and dispersal of seeds especially by small livestock, wild animals and birds ☆ Habitat management especially catering to the needs of livestock production systems of the native farming cultures

A comprehensive attempt of the evaluation of the ecosystem services in the context of the parts of the Himalayan mountains has been made by Singh (2005). All services, of course, cannot be measurable. Most of the services provided by livestock component of the ecosystem that contribute to the services of mountain agro-ecosystems are not measurable.

Evaluating ecosystem services, in fact, is not necessitated for halting or reversing the processes of ecosystem degradation or attaining the state of sustainability. Even the social welfare is not the goal of the evaluation. For the constant flow of services from an ecosystem, ecological integrity of the system is the must. Then what is the need of monitory measurement of this aspect of the situation? Should philosophical gratification be an indispensable ingredient of the human intellect, it can be secured merely by knowing and understanding the elements, processes and complexity that make up a congenial environment to constantly generate and flow services vital for life.

All processes going on in the ecosystems are not regarded to be services for human beings. Some of the services generated out of these ecosystem processes are of potential use for human welfare. The valuation of the services, indeed, is in relation to human species only. Ecosystem services for the welfare and sustenance of other innumerable species on the planet are not in question. These species, seemingly existing on the mercy of human species, are themselves valuated for utilization in the human economic system virtually for human welfare. Symbiosis between human species and the rest of the life forms is not the crux of the thinking behind ecosystem analysis. Ecosystem services criterion merely for human species is opposed to the very concept of symbiosis amidst the ecosystems, species and varieties of species.

The concept of ecosystem valuation promotes the preponderance of one species on Earth, *i.e.*, the human species. The valuation is selective too. It makes gradation in the services to originate from complex ecosystem processes and thus naturally emphasizes what is economically more important for human species. In essence, this is opposed to heterogeneity, which is the rule of life in totality. What is the most pronounced danger for humanity on Earth is not the religious fundamentalism, terrorism, or nuclear catastrophe, but the turning of Earth into human monocultures

(Singh 2005). Monocultures introduce vulnerability in the systems that support them and are ultimately liable to destruction and elimination. Their care, their raising, their protection and short-term existence requires heavy cost to pay. Heterogeneity, on the other hand, depicts considerably high degree of resilience, stability and sustainability.

One thing inherent in the concept of economic valuation is the identification of providers and receivers. These two categories are of the people who should pay the prices and those who should be the gainer. Amongst all sorts of the stakeholders, they might be the communities inhabiting in different areas, such as ecologically more flourishing ones and those facing a state of ecological deprivation. The latter are naturally the receivers of the services and the former are the providers of the services. In terms of the economic gains through ecosystem services the reverse would be the case. This pattern, however, is neither ethical nor practical and, in the setting of Indian cultural ethos, unacceptable also.

More pronounced beneficiaries, however, are not the communities, but the ones ruling over the economic models developed with the eventual aim of globalisation. The rapid processes of globalisation, in fact, are setting the terms and conditions for the maximum possible exploitation of all the natural resources–products as well as services–that are life-producing, life-supporting life-enhancing, and life-sustaining. The concept of ecosystem services' evaluation emanates from the very basic ideology of globalisation. We must pose the question: whose globalisation? Globalisation for whom?

Jodha (1998) identifies three elements which individually or jointly strengthen the ecosystem-social links and contribute to the natural resource-friendly traditional management systems: (a) a total dependence-driven stake in the protection of natural resources; (b) close proximity and a functional knowledge-driven approach to resource use; and (c) local control-determined sanctions and facilities governing resource use. We can derive some modalities for the evaluation of ecosystem services from the ecosystem-social system links as elaborated by Jodha.

Ecosystem services, from the eco-philosophical angle, are meant for the all life partners on Earth, not merely for human species. Human welfare and posterity are inevitably linked with the blossoming of all life forms on Earth. Again, the ecosystem services should be linked with the welfare of entire humanity and they should not be monopolised in the interest of the multinational corporate or in the interest of elite sections of the people. Commoditization of vital ecosystem services is tied with the reductionist thinking, not with the progressive thinking or ecological philosophy.

Therefore, there should be some ethical basis of the valuation of ecosystem services; a basis that must ensure conservation and equitable utilization of resources and promote symbiosis and life-enhancing values. It must embrace reverential attitude towards nature and all processes and services that lead to the affluence of nature and sustainable economies. Ecosystem services' evaluation concept, instead of linking up with global GNP, should articulate in Gross Global Happiness.

References

GRAIN. 2005. Air, don't sell yourself. *Seedling*, April 2005, 34-40.

Hobson, P.N. 1988. The Rumen Microbial Ecosystem. *Elsevier Applied Science*, New York.

http://www.prisma.org.sv/pubs.

Jodha, N.S. 1998. Reviving the Social System-Ecosystem Links in the Himalayas. In Berkes, F.; Folke, C.; and Colding, J. (eds.) *Linking Social and Ecological Systems*. Cambridge: Cambridge University Press.

Kamra, D.N. 2004. Rumen Microbial Ecosystem and Utilisation of Lignocellulosic Crop Residues and Forest Based Fodders: New Tactics of Animal Nutrition. In Singh, V. and Gautam, P.L. (eds.) *Livestock Production Systems for Sustainable Food Security and Livelihoods in Mountain Areas*, 42-50. Pantnagar: GB Pant University of Agriculture and Technology.

Ørskov, E.R. 2005. Animal in Natural Interaction with Soil, Plant and People with emphasis on Asia. In Rowlinson, P; Wachirapakorn, C.; Pakdee, P.; and Wanapat, M. (eds.) *Integrating Livestock-Crop Systems to Meet the Challenges of Globalisation*, Volume 1. British Society of Animal Sciences, U.K.

Share International. 2000. *C & I India Update*, Vol 19, 3, April 2000.

Singh, S.P. 2005. Ecosystem Services. Keynote address to the Workshop on Integrated Management of Natural Resources in Mountains at Pantnagar, 28 Nov.-2 Dec. 2005.

Singh, V. 1998. *Draught Animal Power in Mountain Agriculture: A Study of Perspectives and Issues in Central Himalayas, India.* ICIMOD-MFS Discussion Paper No. 2. Kathmandu: ICIMOD.

Singh, V. 2002. Policy Issues of Marginal Lands and Farms in Uttaranchal, India. Paper presented at the Workshop on *Issues and Options of Marginal Mountain Farms in the HKH* organized by International Centre for Integrated Mountain Development (ICIMOD), Kathmandu, Nepal , July 8-10, 2002.

Singh, V. 2005. Eco-Philosophy, Equity and Sustainable Development: Potential Tools for World Peace. Paper presented at the National Conference on *Role of India in Promoting World Peace* organized by UNESCO and World Peace Centre, Pune on February 3-4, 2005.

Werf, E. van der. 1993. Ecological Farming Principles. In Singh, V. (ed.) *Eco-crisis in the Himalaya: Causes, Consequences and Way Out.* Dehradun: IBD.

Chapter 10

Himalayan Rangelands as a Resource Base of Fodders

☆ *R.D. Gaur and Vir Singh*

Rangeland-livestock relationships are as old as the history of agriculture and animal domestication. In plain areas where most of the rangelands have been transformed into cultivated lands, livestock-rangeland relationships have broken down to a great extent. Here, livestock depend maily on cultivated lands for fodder supplies. In the Himalayan mountains, however, livestock still depend on uncultivated fodders that are supplied from rangelands. Cultivated lands provide only the crop residues. No piece of the cultivated land is devoted to fodder cultivation. Mountain rangelands provide a variety of fodders, and this variety changes from area-to-area and from place-to-place in the same area.

Although fodders species occur plentifully in the rangelands throughout the Himalayan mountains, mid-altitude mountains are especially rich in fodder resources. A list of such species belonging to different plant families is presented in Table 10.1.

Table 10.1: Fodder Species in the mid-altitude Himalayan Rangelands

Family	*Fodder Species*	*Altitude (m)*
ACANTHACEAE	*Adhatoda vasica* Nees	1700
	Aechmanthera gossypina (Nees) Nees	1400
	Barleria cristata L.	1600
	Dicliptera bupleuroides Nees	1800
	Dicliptera bupleuroides Nees	1650
	Justicia simplex (D. Don) Yamazaki	1700
	Strobilanthes atropurpureus Nees	1700
AMARANTHACEAE	*Achyranthes bidentata* Blume	1700
	Alternanthera sessilis L.	1250
	Amaranthus spinosus L.	1400
	Amaranthus viridis L.	1300
	Cyathula tomentosa (Roth) Moq.	1600
	Digera muricata (L.) Martius	1200
	Gomphrena celosioides Martius	1250
	Pupalia lappacea (L.) Juss.	1700
ANACARDIACEAE	*Buchanania latifolia* Roxb.	1450
	Lannea coromandelica (Houttuyn) Merrill	1500
	Rhus semialata Murray	1650
	Rhus wallichii Hook. f.	1700
APIACEAE	*Bupleurum hamiltonii* Balakrishnan	1700
	Pimpinella achilleifolia (DC.) C.B. Clarke	1650
	Pimpinella acuminata (Edgew.) C.B. Clarke	1700
	Pimpinella diversifolia DC.	1700
APOCYNACEAE	*Wrightia tomentosa* (Roxb.) Roemer & Schultes	1450
AQUIFOLIACEAE	*Ilex dipyrena* Wallich	2200
	Ilex excelsa (Wallich) Hook. f.	1500
	Ilex odorata Buch.-Ham. ex D. Don	1500
ARALIACEAE	*Hedera nepalensis* K. Koch	2300
ASTERACEAE	*Anaphalis cinnamomea* C.B. Clarke	1700
	Artemisia nilagirica C.B. Clarke	1700

Contd...

Table 10.1–Contd...

Family	*Fodder Species*	*Altitude (m)*
	Artemisia parviflora Roxb.	1700
	Aster molliusculus Lindley ex DC.) C.B. Clarke	1650
	Bidens biternata (Lour.) Merril & Sherff	1250
	Bidens pilosa L.	1650
	Chrysanthemum leucantheum L.	1650
	Erigeron annuus	1650
	Erigeron bonariensis L.	1650
	Erigeron canadensis L.	1650
	Erigeron karvinskianus DC.	1650
	Eupatorium adenophorum Sprengel	1700
	Galinsoga parviflora Cav.	1650
	Lactuca dissecta D.Don	2250
	Lactuca dissecta D.Don	2250
	Lactuca dolichophylla Kitamura	2250
	Laggera alata (D.Don) Schultz-Bipontinus ex Oliver	2250
	Sonchus brachyotus DC.	1700
	Taraxacum officinale Weber	1650
BALSAMINACEAE	*Impatiens scabrida* DC.	1700
BASELLACEAE	*Basella rubra* L.	1700
BERBERIDACEAE	*Berberis aristata* DC.	2250
	Berberis asiatica Roxb. ex DC.	1650
BETULACEAE	*Betula alnoides* Buch.-Ham. ex D. Don	2200
BURSERACEAE	*Boswellia serrata* Roxb. ex Colebr.	1250
	Garuga pinnata Roxb.	1200
CAESALPINIACEAE	*Bauhinia acuminata* L.	1250
	Bauhinia purpurea L.	1250
	Bauhinia racemosa Lam.	1400
	Bauhinia vahlii Wight & Arn.	1650
	Bauhinia variegata L.	1200
	Caesalpinia bonduc (L.) Roxb.	1200

Contd...

Table 10.1–Contd...

Family	Fodder Species	Altitude (m)
	Caesalpinia decapetala (Roth) Alston	1400
	Cassia absus L.	1000
	Cassia mimosoides L.	1650
	Cassia tora L.	1650
	Bauhinia semla Wunderlin	800
CAMPANULACEAE	*Cyananthus lobatus* Wallich ex Benth	3000
CAPPARACEAE	*Capparis sepiaria* L.	1200
	Capparis zeylanica L.	800
	Crataeva adansonii DC.	500
CAPRIFOLIACEAE	*Lonicera quinquelocularis* Hardwicke	1700
	Viburnum coriaceum Blume	1700
CARYOPHYLLACEAE	*Drymaria cordata* (L.) Willd. ex Roemer & Schultes	1650
	Stellaria media (L.) Villars	2000
CELASTRACEAE	*Euonymus echinatus* Wallich	2200
	Euonymus hamiltonianus Wallich	2500
	Euonymus pendulus Wallich	2200
	Euonymus tingens Wallich	2400
	Euonymus tingens Wallich	1800
CHENOPODIACEAE	*Chenopodium album* L.	1400
	Chenopodium ambrosioides L.	1550
	Chenopodium botrys L.	2200
	Chenopodium foliosum (Moench) Ascherson	2300
	Chenopodium murale L.	1000
COMBRETACEAE	*Anogeissus latifolius* (Roxb. ex DC.) Wallich ex Richard	1400
	Terminalia alata Heyne ex Roth	1500
	Terminalia chebula Retz.	1600
COMMELINACEAE	*Commelina benghalensis* L.	1700
CONVOLVULACEAE	*Cuscuta reflexa* Roxb.	1700
	Ipomoea muricata (L.) Jacquin	1650

Contd...

Table 10.1–Contd...

Family	*Fodder Species*	*Altitude (m)*
CORIARIACEAE	*Coriaria nepalensis* Wallich	2300
CORNACEAE	*Cornus oblonga* Wallich	1700
	Swida macrophylla (Wallich) Sojak	2250
	Swida oblonga (Wallich) Sojak	1700
CORYLACEAE	*Carpinus viminea* Lindley	2300
CUCURBITACEAE	*Gymnostemma pedata* Blume	1650
	Trichosanthes palmata Roxb.	1700
CYPERACEAE	*Carex cruciata* Wahlenberg	1650
	Cyperus corymbosus Rottboell	1350
	Cyperus monocephalus Roxb.	1650
	Cyperus nutans Vahl	1650
	Cyperus rotundus L.	1700
DILLENIACEAE	*Dillenia indica* L.	1200
DIOSCOREACEAE	*Dioscorea belophylla* (Prain) J.O. Voigt ex Haines	1200
	Dioscorea bulbifera L.	1650
	Dioscorea deltoidea Wallich ex Grisebach	2200
	Dioscorea glabra Roxb.	1000
	Dioscorea hispida Dennstaedt	800
	Dioscorea melanophyma Prain & Burkill	1500
	Dioscorea pentaphylla L.	1500
	Dioscorea pubera Blume	600
EBENACEAE	*Diospyros montana* Roxb.	1250
	Diospyros tomentosa Roxb.	1200
EHRETIACEAE	*Ehretia acuminate* R.Br.	1200
	Ehretia laevis Roxb.	1500
ELAEAGNACEAE	*Elaeagnus parviflora* Wallich ex Royle	2600
	Hippophae salicifolia D. Don	2200
EUPHORBIACEAE	*Bridelia retusa* (L.) Sprengel	1000
	Bridelia verrucosa Haines	1000
	Euphorbia pilosa L	2400
	Euphorbia chamaesyce L.	1650

Contd...

Table 10.1–Contd...

Family	*Fodder Species*	*Altitude (m)*
	Euphorbia hirta L.	1200
	Euphorbia hypericifolia L.	1200
	Phyllanthus parvifolius Buch.-Ham. ex D. Don	2000
FABACEAE	*Alysicarpus bupleurifolius* (L.) DC.	1000
	Alysicarpus heyneanus Wight & Arn.	1000
	Alysicarpus procumbens (Roxb.) Schindler	400
	Alysicarpus vaginalis (L.) DC.	1000
	Argyrolobium roseum (Cambess.) Jaubert & Spach	1600
	Astragalus chlorostachys Lindley	3000
	Astragalus graveolens Buch.-Ham. ex Benth.	2400
	Astragalus leucocephalus Graham ex Benth.	2200
	Butea monosperma (Lam.) Kuntze	400
	Codariocalyx motorius (Houttuyn) Ohashi	1200
	Crotalaria medicaginea Lam.	1400
	Dalbergia lanceolata L. f.	500
	Dalbergia latifolia Roxb.	600
	Dalbergia sericea G. Don	600
	Dalbergia sissoo Roxb.	1200
	Dalbergia volubilis Roxb.	1000
	Erythrina variegata L.	800
	Flemingia bracteata (Roxb.) Wight	1200
	Flemingia macrophylla (Willd.) Prain ex Merrill	1200
	Flemingia paniculata Wallich ex Benth.	800
	Flemingia procumbens Roxb.	1600
	Flemingia semialata Roxb. ex Aiton f.	1700
	Flemingia strobilifera (L.) R.Br.	1700
	Indigofera dosua Buch.-Ham. ex D.Don	1700
	Indigofera heterantha Wallich ex Brandis	1250

Contd...

Table 10.1–Contd...

Family	*Fodder Species*	*Altitude (m)*
	Lathyrus aphaca L.	1200
	Lathyrus sativus L.	1200
	Lathyrus sphaericus Retz.	1000
	Lespedeza gerardiana Graham ex Maxim.	2250
	Lespedeza juncea (L.f.) Persoon	2250
	Lotus corniculata L.	2600
	Medicago lupilina L.	1200
	Medicago polymorpha L.	1200
	Melilotus alba Medikus ex Desrousseaux	1250
	Melilotus indica (L.) Allioni	1500
	Microtyloma sar-garhwalensis Gaur and Dangwal	1800
	Milletia extensa (Benth.) Baker	1000
	Milletia extensa (Benth.) Baker	1600
	Mucuna nigricans (Lour.) Steudel	1250
	Mucuna pruriens (L.) DC.	1200
	Ougeinia dalbergioides Benth.	1650
	Phyllodium pulchellum (L.) Desvaux	1000
	Pongamia pinnata (L.) Pierre	1250
	Pterocarpus marsupium Roxb.	400
	Pueraria tuberose (Roxb. ex Willd.) DC.	1200
	Rhynchosia minima (L.) DC.	1200
	Sesbania sesban (L.) Merrill	1000
	Shuteria involucrate (Wallich) Wight & Arn.	800
	Shuteria vestita Wight & Arn.	1500
	Smithia ciliate Royle	1000
	Spatholobus parviflorus (Roxb. ex DC.) Kuntze	600
	Tephrosia candida (Roxb.) DC.	1200
	Tephrosia pumila (Lam.) Persoon	500
	Tephrosia purpurea (L.) Persoon	800
	Trifolium alexandrinum L.	1200

Contd...

Table 10.1–Contd...

Family	Fodder Species	Altitude (m)
	Trifolium repens L.	1650
	Vicia sativa L.	1200
	Vicia tenera Graham ex Benth.	1800
	Vigna aconitifolia (Jacquin) Marechal	1200
	Vigna sublobata (Roxb.) Babu & Sharma	1600
	Vigna trilobata (L.) Verdc.	800
	Vigna vexillata (L.) A. Richard	1200
	Desmodium elegans DC.	1800
	Desmodium laxiflorum DC.	1650
	Desmodium microphyllum (Thunb.) DC.	1800
	Desmodium multiflorum DC.	1800
	Desmodium triflorum (L.) DC.	1800
	Glycine max (L.) Merrill	1600
	Indigofera cassioides Rottler ex DC.	1600
FAGACEAE	*Castanopsis tribuloides* (J.E. Smith) A. DC.	2000
	Quercus floribunda Lindley ex Rehder	2500
	Quercus glauca Thunb.	2000
	Quercus leucotrichophora A. Camus	1650
	Quercus semecarpifolia J.E. Smith	2000
FLACOURTIACEAE	*Flacourtia indica* (Burm. f.) Merrill	1200
GERANIACEAE	*Geranium wallichianum* D. Don ex Sweet	1700
	Swertia angustifolia Buch.-Ham. ex D. Don	1700
GESNERIACEAE	*Didymocarpus pedicellatus* R.Br.	1250
HIPPOCASTANACEAE	*Aesculus indica* (Colebr. ex Cambess.) Hook.	2200
HYDRANGEACEAE	*Deutzia staminea* R.Br. ex Wallich	1500
HYPERICACEAE	*Hypericum choisianum* Wallich ex N. Robson var. *H. hookerianum* Wight & Arn.	2000
	Hypericum elodeoides Choisy	2400
	Hypericum japonicum Thunb. ex Murray	2300
	Hypericum oblongifolium Choisy	2300
	Hypericum perforatum L.	2000

Contd...

Table 10.1–Contd...

Family	*Fodder Species*	*Altitude (m)*
	Hypericum uralum Buch.-Ham. ex D. Don	2000
	Hypericum wightianum Wallich ex Wight & Arn.	2500
LAMIACEAE	*Leucas lanata* Benth.	1650
	Origanum vulgare L.	1700
	Plectranthus japonicus (Burm. f.) Koidz	1700
	Salvia hians Royle ex Benth	2250
	Salvia leucantha Cav.	1650
	Thymus biflorus Buch.-Ham. ex D.Don	1700
LAMINACEAE	*Leucas indica* (L.) R.Br. ex Vatke	1650
LAURACEAE	*Litsea elongate* (Nees) Hook. f.	1450
	Cinnamomum tamala (Buch.-Ham.) Nees & Ebermaeir	1700
	Dodecadenia grandiflora Nees	2500
	Litsea glutinosa (Lour.) Robinson	1200
	Litsea monopetala (Roxb.) Persoon	1200
	Machilus duthiei King ex Hook. f.	1650
	Neolitsea cuipala (Buch.-Ham. ex D. Don) Kostermans	2300
	Nitsea monopetala (Roxb.) Persoon	1200
	Persea duthiei (King ex Hook. f.) Kostermans	2200
	Phoebe laceolata (Nees) Nees	1200
LORANTHACEAE	*Scurrula cordifolia* (Wallich) G. Don	400
	Scurrula elata (Edgew.) Danser	2400
MAGNOLIACEAE	*Miliusa velutina* (Dunal) Hook. f. & Thomsom	700
MALVACEAE	*Hibiscus cannabinus* L.	800
	Kydia calycina Roxb.	1200
	Urena lobata L.	1650
MELIACEAE	*Melia azedarach* L.	1300
	Toona ciliata Roemer	1650
	Toona serrata (Royle) M. Roemer	1800

Contd...

Table 10.1–Contd...

Family	*Fodder Species*	*Altitude (m)*
MENISPERMACEAE	*Cissampelos pareira* L.	1200
	Tinospora sinensis (Lour.) Merrill	1650
MIMOSACEAE	*Acacia catechu* (L.f.) Willd.	1250
	Acacia farnesiana (L.) Willd.	1000
	Albizia chinensis (Osbeck) Merrill	600
	Albizia julibrissin Durazzini	1600
	Albizia lebbeck (L.) Benth.	1400
	Albizia lucidior (Steudel) Nielsen	800
	Albizia odoratissima (L.f.) Benth.	800
	Albizia procera (Roxb.) Benth.	600
	Mimosa himalayana Gamble	1600
MORACEAE	*Ficus arnottiana* (Miq.) Miq.	600
	Ficus auriculata Lour.	1000
	Ficus benghalensis L.	1200
	Ficus clavata Wallich ex Miq.	1550
	Ficus cunia Buch.-Ham, ex Roxb.	1550
	Ficus glaberrima Blume	1000
	Ficus hederacea Roxb.	1800
	Ficus hipsida L.f.	1000
	Ficus laminose Hardwicke	600
	Ficus neriifolia Smith var. *Ficus nemoralis* Wallich ex Miq.	1550
	Ficus palmata Forsk.	1550
	Ficus racemosa L.	900
	Ficus religiosa L.	1200
	Ficus roxburghii Wallich ex Miq.	1550
	Ficus rumphii Blume	1200
	Ficus sarmentosa Buch.-Ham. ex J.E. Smith	2000
	Ficus semicordata Buch.-Ham. ex J.E. Smith	1400
	Morus alba L.	1650

Contd...

Table 10.1–Contd...

Family	*Fodder Species*	*Altitude (m)*
	Morus macroura Miq.	1200
	Morus serrata Roxb.	1200
	Streblus asper Lour.	600
MYRSINACEAE	*Myrsine africana* L.	1700
	Myrsine semiserrata Wallich	2000
OLEACEAE	*Fraxinus micrantha* Lingelsheim	1650
	Ligustrum compactum (Wallich ex DC.) Hook. f. & Thomson ex Brandis	1750
	Ligustrum indicum (Lour.) Merrill	1750
	Olea glandulifera Wallich ex G.Don	1700
OROBANCHACEAE	*Orobanche aegyptiaca* Persoon	400
OXALIDACEAE	*Oxalis corniculata* L.	1650
	Oxalis dehradunensis Raizada	1650
PASSIFLORACEAE	*Passiflora caerulea* L.	1650
PLANTAGINACEAE	*Plantago lanceolata* L.	1700
POACEAE	*Agropyron longe-aristatum* Boiss.	1700
	Agrostis nervosa Nees ex Trinius	2300
	Agrostis pilosula Trinius	1650
	Agrostis vinealis Schreber	1250
	Alloteropsis cimicina (L.) Stapf	1400
	Andropogon munroi C.B. Clarke	2400
	Apluda aristata L.	1200
	Apluda mutica L.	1650
	Aristida adscensionis L.	1200
	Arthraxon hispidus (Thumb.) Makino	2200
	Arthraxon lanceolatus (Roxb.) Hochst.	1550
	Arthraxon lancifolius (Trinius) Hochst.	1600
	Arthraxon lancifolius (Trinus) Hochst.	1650
	Arthraxon prionodes (Steudel) Dandy	2300
	Arthraxon prionodes (Steudel) Dandys	1700
	Arundinella bengalensis (Sprengel) Druce	1500
	Arundinella birmanica Hook. f.	1600

Contd...

Table 10.1–Contd...

Family	Fodder Species	Altitude (m)
	Arundinella nepalensis Trinius	1650
	Arundinella nervosa (Roxb.) Nees ex Hook. & Arn.	2300
	Arundinella pumila (Hochst. ex Richard)	1600
	Arundo donax L.	2300
	Avena fatua L.	1650
	Bambusa arundinacea Willd.	1200
	Bothriochloa foulkesii (Hook.f.) Henrard	1400
	Bothriochloa glabra (Roxb.) A. Camus	1800
	Bothriochloa ischaemum (L.) Keng	1200
	Brachiaria distachya (L.) Stapf	1000
	Brachiaria ramosa (L.) Stapf	1200
	Brachiaria villosa (Lam.) A. Camus	1600
	Brachypodium sylvaticum (Hudson) P. Beauv.	1650
	Bromus catharticus Vahl	1600
	Capillipedium assimile (Steudel) A. Camus	1200
	Capillipedium assimile Steudel	1700
	Capillipedium huegelii (Hackel) Stapf	1600
	Capillipedium parviflorum (R.Br.) Stapf	1650
	Capillipedium parviflorum (R.Br.) Stapf	2400
	Cenchrus ciliaris L.	400
	Chloris dolichostachya Lagasca	600
	Chrysopogon aciculatus (Retz.) Trinius	800
	Chrysopogon fulvus (Sprengel) Chiovenda	1000
	Chrysopogon gryllus (L.) Trinius	1650
	Coix lacryma-jobi L.	1000
	Cymbopogon flexuosus (Nees ex Steudel) W. Watson	600
	Cymbopogon jwarancusa (Jones) Schultes	1000
	Cymbopogon martinii (Roxb.) W. Watson	1100
	Cynodon dactylon (L.) Persoon	1650

Contd...

Table 10.1–Contd...

Family	*Fodder Species*	*Altitude (m)*
	Cyrtococcum accrescens (Trinius) Stapf	1400
	Dactyloctenium aegypticum (L.) P. Beauv.	1650
	Dendrocalamus strictus (Rox.) Nees	1200
	Desmostachya bipinnata (L.) Stapf	400
	Dichanthium annulatum (Forsk.) Stapf	1700
	Digitaria abludens (Roemer & Schultes) Veldkamp	1200
	Digitaria adscendens (Kunth) Hennard	1500
	Digitaria bicornis (Lam.) Roemer & Schultes	500
	Digitaria ciliaris (Retz.) Koeler	1200
	Digitaria cruciata (Nees ex Steudel) A. Camus	1650
	Digitaria granularis (Trinius) Henrard	1650
	Digitaria setigera Roth ex Roemer & Schultes	700
	Digitaria stricta Roth ex Roemer & Schultes	1650
	Echinochloa crus-galli (L.) P. Beauv.	1650
	Echinochloa frumentacea Link	1650
	Eleusine compressa (Forsk.) Ascherson & Schweinfurth ex Christensen	1000
	Eleusine coracana (L.) Gaertner	1550
	Eleusine indica (L.) Gaertner	1200
	Eleusine indica (L.) Gaertner	1650
	Eragrostiella nardoides (Trinius) Bor	1200
	Eragrostis atrovirens (Desfontaines) Trinius ex Steudel	1650
	Eragrostis japonica (Thumb.) Trinius	1450
	Eragrostis minor Host	800
	Eragrostis nigra Nees ex Steudel	2200
	Eragrostis pilosa (L.) P. Beauv.	800
	Eragrostis tenella (L.) P. Beauv. ex Roemer & Schultes	600
	Eragrostis unioloides (Retz.) Nees ex Steudel	1200

Contd...

Table 10.1–Contd...

Family	Fodder Species	Altitude (m)
	Eulalia hirtifolia (Hackel) A. Camus	1400
	Eulalia leschenaultiana (Decne.) Ohwi	1000
	Eulalia mollis (Grisebach) Kuntze	2500
	Eulalia quadrinervis (Hackel) Kuntze	800
	Eulaliopsis binata (Retz.) Hubbard	1200
	Festuca gigantea (L.) Villars	2500
	Festuca rubra L.	1200
	Garnotia tenella (Arn. ex Miq.) Janowsky	1500
	Hackelochloa granularis (L.) Kuntze	1200
	Helictotrichon virescens (Nees ex Steudel) Henrard	2400
	Hemarthria compressa (L.f.) R.Br.	1000
	Heteropogon contortus (L.) P. Beauv. ex Roemer & Schultes	1450
	Imperata cylindrica (L.) P. Beauv.	1700
	Isachne albens Trinius	1700
	Ischaemum rugosum Salisbury	400
	Koeleria macrantha (Ledebour) Schultes	1500
	Leptochloa chinensis (L.) Nees	400
	Leptochloa panacea (Retz.) Ohwi	1200
	Lolium temulentum L.	1000
	Microstegium falconeri (Hook. f.) Clayton	1600
	Microstegium nudum (Trinius) A. Camus	800
	Miscanthus nepalensis (Trinius) Hackel	1700
	Mnesithea laevis (Retz.) Kunth	1200
	Muhlenbergia huegelii Trinius	1650
	Neyraudia arundinacea (L.) Hennard	1000
	Oplismenus burmannii (Retz.) P. Beauv.	600
	Oplismenus compositus (L.) P. Beauv.	1700
	Oplismenus undulatifolius (Ardoino) P. Beauv.	2300
	Oropetium thomaeum (L.f.) Trinius	1000

Contd...

Table 10.1–Contd...

Family	*Fodder Species*	*Altitude (m)*
	Panicum antidotale Retz.	1450
	Panicum curviflorum Hornem.	1000
	Panicum miliaceum L.	1650
	Panicum paludosum Roxb.	1600
	Panicum psilopodium Trinius	1700
	Paspalidium flavidum (Retz.) Camus	1650
	Paspalum paspalodes (Michaux) Scribn.	1650
	Paspalum scrobiculatum L.	1400
	Pennisetum flaccidum Grisebach	1650
	Pennisetum orientale Richard	1500
	Pennisetum typhoides Burman	1650
	Perotis hordeiformis Nees ex Hook & Arn.	600
	Phacelurus speciosus (Steudel) C.E. Hubbard	2400
	Phalaris minor Retz.	1500
	Phragmites karka (Retz.) Trinius ex Steudel	400
	Poa annua L.	1650
	Poa nemoralis L.	2400
	Poa pratensis L.	1850
	Pogonatherum crinitum (Thunb.)	1200
	Pogonatherum paniceum (Lam.) Hackel	2200
	Pollinia quadrinervis Hackel	1650
	Polypogon fugax Nees ex Steudel	1200
	Polypogon monspeliensis (L.) Desfontaines	1200
	Polypogon monspeliensis (L.) Desfontaines	1250
	Pseudosorghum fasciculare (Roxb.) A. Camus	600
	Rottboellia cochinchinensis (Lour.) W.D. Clayton	400
	Saccharum bengalensis Retz.	600
	Saccharum filifolium Steud.	1500

Contd...

Table 10.1–Contd...

Family	Fodder Species	Altitude (m)
	Saccharum longisetosum (Andersson ex Benth.) Narayanswami ex Bor	1000
	Saccharum narenga (Nees ex Steudel) Hackel	600
	Saccharum rufipilum Steudel	1650
	Saccharum spontaneum L.	1200
	Saccharum spontaneum L.	800
	Sacciolepsis indica (L.) A. Chase	2000
	Setaria barbata (Lam.) Kunth	600
	Setaria glauca (L.) P. Beauv.	1650
	Setaria homonyma (Steudel) Choivenda	1650
	Setaria intermedia Roemer & Schultes	1200
	Setaria pumila (Poiret) Roemer	1650
	Setaria verticillata (L.) P. Beauv.	1200
	Setaria viridis (L.) P. Beauv.	2500
	Sinarundinaria anceps (Mitf.) Chao & Renvoize	2250
	Sinarundinaria falcate (Nees) Chao & Renvoize	2250
	Sorghum halepense (L.) Persoon	600
	Sorghum nitidum (Vahl) Persoon	300
	Sporobolus diander (Retz.) P. Beauv.	1500
	Sporobolus fertilis (Steudel) Clayton	1200
	Sporobolus piliferus (Trinius) Kunth	1950
	Sporobolus spicatus (Vahl) Kunth	1800
	Stipa roylei (Nees) Mez	2300
	Stipa sibirica (L.) Lam.	2300
	Thamnocalamus falconeri Hook. f. ex Munro	2200
	Thamnocalamus spathiflora (Trinius) Munro	2300
	Themeda anathera (Nees ex Steudel) Hackel	1650

Contd...

Table 10.1–Contd...

Family	*Fodder Species*	*Altitude (m)*
	Themeda arundenacea (Roxb.) Ridley	1200
	Themeda triandra Forsk	2000
	Themeda villosa (Poiret) A. Camus	600
	Thysanolaena maxima (Roxb.) Kuntze	1200
	Tragus roxburghii Panigrahi	600
	Tripogon filiformis Nees ex Steudel	1650
	Urochloa panicoides P. Beauv.	800
	Vetiveria zizanioides (L.) Nash	600
POLYGONACEAE	*Aconogonum molle* (D. Don) Hara	1300
	Fagopyrum dibotrys (D.Don) Hara	1700
	Fagopyrum esculentum (L.) Moench	1500
	Polygonum capitatum Buch.-Ham. ex D.Don	1650
	Polygonum hydropiper L.	1650
	Rumex dentatus L.	1200
	Rumex hastatus D. Don	1650
RANUNCULACEAE	*Anemone vitifolia* Buch.-Ham. ex DC.	2400
	Clematis barbellata Edgew	2600
	Clematis buchananiana DC.	1650
	Clematis connata DC.	2500
	Clematis grata Wallich	2700
	Thalictrum foliolosum DC.	1650
RHAMNACEAE	*Berchemia floribunda* (Wallich) Brongniart	1200
	Rhamnus persica Boissiers	1200
	Rhamnus purpureus Edgew.	1500
	Ziziphus mauritiana Lam.	600
	Ziziphus nummularia (Burm. f.) Wight & Arn.	400
	Ziziphus xylopyrus (Retz.) Willd.	400
ROSACEAE		
	Pyrus pashia Buch.-Ham. ex D. Don	1400
	Agrimonia pilosa Ledebour	1650
	Cotoneaster lindleyi Steudel	2600
	Prunus armeniaca L.	1650

Table 10.1–Contd...

Family	*Fodder Species*	*Altitude (m)*
	Prunus cerasoides D. Don	1500
	Prunus cornuta (Wallich ex Royle) Steudel	2400
	Rubus ellipticus Smith	1650
	Rubus paniculatus Smith	1650
RUBIACEAE	*Adina cordifolia* (Roxb.) Hook. f. ex Brandis	1500
	Catunaregam uliginosa (Retz.) Sivarajan	400
	Galium aparine L.	1550
	Galium elegans Wallich	1600
	Galium rotundifolium L.	1700
	Hymenodictyon flaccidum Wallich	1200
	Hymenodictyon orixense (Roxb.) Mabberley var. *Hymenodictyon excelsum* (Roxb.) Wallich	800
	Leptodermis lanceolata Wallich	1700
	Mitragyna parviflora (Roxb.) Korthals	800
	Pavetta indica L.	1000
	Rubia cordifolia non L.	1700
	Rubia manjith Roxb. ex Fleming	1400
	Spermadictyon sauveolens Roxb.	2000
RUTACEAE	*Boenninghausenia albiflora* (Hook.) Reichb. ex Meisn.	1700
SABIACEAE	*Meliosma dilleniifolia* (Wallich ex Wight & Arn.) Walpers	2600
SALICACEAE	*Populus ciliata* Wallich ex Royle	1200
	Salix acemophylla Boissier	1400
	Salix acutifolia Willd.	2400
	Salix denticulata Anderson	2600
	Salix disperma Roxb. ex D. Don	2600
	Salix tetrasperma Roxb.	1000
SAPINDACEAE	*Pistacia integerrima* Stewart	1700
	Schleichera oleosa (Lour.) Oken	1200
SAPOTACEAE	*Madhuca longifolia* (Koenig) Mac Bride	1250
SAURAUIACEAE	*Saurauia nepaulensis* DC.	1600

Contd...

Table 10.1–Contd...

Family	*Fodder Species*	*Altitude (m)*
SAXIFRAGACEAE	*Astilbe rivularis* Buch.-Ham. ex D.Don	1950
SCROPHULARIACEAE	*Lindenbergia indica* (L.) Vatke	1700
	Torenia cordifolia Roxb.	1700
SMILACACEAE	*Smilax aspera* L.	1650
	Smilax glaucophylla Klotzsch	2200
SOLANACEAE	*Nicandra physalodes* (L.) Gaertner	1700
	Solanum nigrum L.	1650
SYMPLOCACEAE	*Symplocos paniculata* (Thumb.) Miq.	2400
	Symplocos racemosa Roxb.	1000
	Symplocos ramosissima Wallich ex G. Don	2500
THEACEAE	*Eurya acuminata* DC.	1700
TILIACEAE	*Grewia asiatica* L.	800
	Grewia eriocarpa A.L. Juss.	800
	Grewia optiva J.R. Drummond ex Burret	1650
	Grewia sapida Roxb. ex DC.	600
	Grewia sclerophylla Roxb. ex D. Don	600
	Grewia serrulata DC.	400
ULMACEAE	*Holoptelea integrifolia* (Roxb.) Planchon	1250
	Trema orientalis (L.) Blume	1200
	Trema politoria (Planchon) Blume	1200
	Ulmus wallichiana Planchon	2500
URTICACEAE	*Boehmeria macrophylla* D.Don	1200
	Boehmeria platyphylla D.Don	1650
	Boehmeria rugulosa Wedd.	1700
	Celtis australis L.	1650
	Debregeasia longifolia (Burm. f.) Wedd.	1450
	Debregeasia salicifolia (D. Don) Rendle	1600
	Girardinia diversifolia (Link) Friis	1800
	Maoutia puya (Hook.) Wedd.	1400
	Maoutia puya (Hook.) Wedd.	1200
	Orecocnide frutescens (Thumb.) Miq.	1000

Contd...

Table 10.1--Contd...

Family	Fodder Species	Altitude (m)
	Pouzolzia zeylanica (L.) J. Bennett & Brown	1500
	Urtica dioica L.	2250
	Urtica parviflora Roxb.	1650
VERBENACEAE	*Caryopteris foetida* (D. Don) Thellung	2300
	Caryopteris odorata (D. Don) B.L. Robinson	1200
	Gmelina arborea Roxb.	1500
VIOLACEAE	*Viola biflora* L.	2700
VITACEAE	*Ampelocissus divaricata* (Wallich ex Lawson) Planchon	1650
	Ampelocissus rugosa (Wallich) Planchon	2000
	Cayratia trifolia (L.) Domin	1250
	Cissus repanda Vahl	1200
	Parthenocissus semicordata (Wallich) Planchon	1700
	Tetrastigma lanceolarium (Roxb.) Planchon	1200
	Vitis flexuosa Thunb.	2200
ZINGIBERACEAE	*Hedychium spicatum* Buch.-Ham. ex J.E. Smith	1700

As is revealed from Table 10.1, as many as 541 fodder species belonging to 83 families of the plants have been found occurring in the mid-altitude Himalayan rangelands. Family Poaceae alone offers as many as 164 species of fodder value for livestock. Fabaceae comprises 69 fodder species, Moraceae 21, Asteraceae 19, Rubiaceae and Urticaceae 13 each, Caeselpiniaceae 11, Lauraceae 10, Mimosaceae 9, and Amaranthaceae, Dioscoreaceae, Hypericaceae and Rosaceae 8 each. Other families have fewer species of fodder value (Table 10.2).

Some wild plants also occur on the vertical stairs of the mountain fields or on margins. These wild plants play an important ecological role in maintaining the fertility of the soil. These plants, according to Gaur (1999), form an intermittent strip in between two crop fields, balancing effects of cultigens. These plants die and decay and thus their nutrients are leached into the terraced fields on lower side. This adds to soil fertility of the terraced fields. Some of the important plants are (Gaur 1999):

Table 10.2: Number of Fodder Species Belonging to Different Families in the Himalayan Rangelands

Sl.No.	*Family*	*Number of Species*
1.	Acanthaceae	7
2.	Amaranthaceae	8
3.	Anacardiaceae	4
4.	Apiaceae	4
5.	Apocynaceae	1
6.	Aquifoliaceae	3
7.	Araliaceae	1
8.	Asteraceae	19
9.	Balsaminaceae	1
10.	Basellaceae	1
11.	Berberidaceae	2
12.	Betulaceae	1
13.	Burseraceae	2
14.	Caeselpiniaceae	11
15.	Campanulaceae	1
16.	Capparaceae	3
17.	Caprifoliaceae	2
18.	Caryophyllaceae	2
19.	Celastraceae	5
20.	Chenopodiaceae	5
21.	Combretaceae	3
22.	Commelinaceae	1
23.	Convolvulaceae	2
24.	Coriariaceae	1
25.	Cornaceae	3
26.	Corylaceae	1
27.	Cucurbitaceae	2

Contd...

Table 10.2–Contd...

Sl.No.	Family	Number of Species
28.	Cyperaceae	5
29.	Dilleniaceae	1
30.	Dioscoreaceae	8
31.	Ebenaceae	2
32.	Ehretiaceae	2
33.	Elaeagnaceae	2
34.	Euphorbiaceae	7
35.	Fabaceae	69
36.	Fagaceae	5
37.	Flacourtiaceae	1
38.	Geraniaceae	1
39.	Gesneriaceae	1
40.	Hippocastanaceae	1
41.	Hydrangeaceae	1
42.	Hypericaceae	8
43.	Lamiaceae	6
44.	Laminaceae	1
45.	Lauraceae	10
46.	Loranthaceae	2
47.	Magnoliaceae	1
48.	Malvaceae	3
49.	Meliaceae	3
50.	Menispermaceae	2
51.	Mimosaceae	9
52.	Moraceae	21
53.	Myrsinaceae	2
54.	Oleaceae	4
55.	Orobanchaceae	1
56.	Oxalidaceae	2
57.	Passifloraceae	1
58.	Plantaginaceae	1

Contd...

Table 10.2–Contd...

Sl.No.	*Family*	*Number of Species*
59.	Poaceae	164
60.	Polygonaceae	7
61.	Ranunculaceae	6
62.	Rhamnaceae	6
63.	Rosaceae	8
64.	Rubiaceae	13
65.	Rutaceae	1
66.	Sabiaceae	1
67.	Salicaceae	6
68.	Sapindaceae	2
69.	Sapotaceae	1
70.	Saurauiaceae	1
71.	Saxifragaceae	1
72.	Scrophulariaceae	2
73.	Smilacaceae	2
74.	Solanaceae	2
75.	Symplocaceae	3
76.	Theaceae	1
77.	Tiliaceae	6
78.	Ulmaceae	4
79.	Urticaceae	13
80.	Verbenaceae	3
81.	Violaceae	1
82.	Vitaceae	7
83.	Zingiberaceae	1

Source: Based on Gaur (1999), Singh (2007) and Singh *et al.* (2008).

Artemisia nilagirica, Arthraxon lanceolatus, Arthraxon prionodes, Arundinella bengalensis, Arundinella nepalensis, Barleria cristata, Bidens biternata, Bidens pilosa, Boehmeria platyphylla, Bupleurum hamiltonii, Capillipedium parviflorum, Crotalaria albida, Cyperus corymbosus, Cyperus monocephalus, Cyperus nutans, Cyperus rotundus, Desmodium laxiflorum, Desmodium microphyllum, Digitaria adscendens, Erigeron bonariensis, Galium aparine, Galium elegans, Girardinia diversifolia, Imperata cylindrica, Leptodermis lanceolata, Oplismenus compositus, Panicum antidotale, Panicum miliaceum, Panicum psilopodium, Pennisetum orientale, Polygonum capitatum, Polygonum hydropiper, Pouzolzia zeylanica, Rubia manjith, Rubus ellipticus, Rumex dentatus, Rumex hastatus, Saccharum bengalensis, Saccharum spontaneum, Salvia hians, Spermadictyon sauveolens, Thalictrum foliolosum, Themeda anathera, Urtica dioica, etc.

Gaur (1999) has brought out comprehensive account of the biodiversity present in the Garhwal district of Uttarakhand. Rangelands serve as rich repositories of fodder plants. As such, these areas are crucial for livestock production, which is the major occupation of mountain farming communities and of a quite sizeable population of nomadic or transhumant people in the Himalayan mountains.

Biodiversity in the rangelands is influenced considerably by altitude and slope orientation. Grasslands at lower and mid-altitudes are often interspersed by trees and shrubs. Amatya (1990), Bhatt and Rawat (1993), Bana and Singh (2004), Singh and Gaur (2005), Singh and Bohra (2005, 2006), Bohra (2006), Singh (2007) and Singh *et al.* (2008) have identified and documented an impressively large number of fodder species occurring in the rangeland ecosystems in parts of the HKH region. Prevalence of individual species is mainly according to altitude and slope orientation.

The Himalayan rangelands truly serve as a rich repository of a variety of uncultivated fodders. Furthermore, these rangelends serve as a rich resource for livestock based livelihoods in the mountains. Rangelands play crucial role in trapping atmospheric carbon and sequester the same into ecosystems. They have enormous bearing on cultivated areas and are critical for maintaining ecological balance in the fragile Himalayan mountains.

References

Amatya, S.M. 1990. *Fodder Trees and their Lopping Cycle in Nepal.* Technical Report, Kathmandu.

Bana, O.P.S. and Singh, V. 2004. Silvi-pasture systems for livestock production in the Hindu Kush-Himalayan region: issues and options. In Singh, V. and Gautam, P.L. (eds.) *Livestock Production Systems for Sustainable Food Security and Livelihoods in Mountain*

Areas. GB Pant University of Agriculture and Technology, Pantnagar.

Bhatt, A.B. and Rawat, N. 1993. Fodder resources of Garhwal: a search for non conventional fodder species. In Rajwar, G.S. (ed.) *Garhwal Himalaya: Ecology and Environment*. pp. 227-239. Ashish Publishing House, New Delhi.

Bohra, B. 2006. *Dairy Farming and Rangeland Resources in Mountain Agro-ecosystems in Uttaranchal*. Ph.D. Thesis. GB Pant University of Agriculture and Technology, Pantnagar.

Gaur, R.D. 1999. *Flora of the District Garhwal, North West Himalaya (with Ethnobotanical Notes)*. Transmedia, Srinagar, Garhwal.

Singh, V. 2007. *Studies on the Rangeland Management of Selected Areas in Kumaun Himalaya*. Ph.D. Thesis. HNB Garhwal University, Srinagar, Uttarakhand, India.

Singh, V. and Bohra, B. 2005. Livestock feed resources and feeding practices in hill farming systems: a review. *Indian Journal of Animal Sciences*, 75: 121-127.

Singh, V. and Bohra, B. 2006. *Dairy Farming in Mountain Areas*. Daya Publishers, New Delhi.

Singh, V. and Gaur, R.D. 2005. The Himalayan rangelands: ecosystem services and ecotourism opportunities. In Rajwar, G.S., Bisht, H., Sharma, Y.K., Kushwaha, M.D., Goswami, D.C. and Rawat, M.S. (eds.) *Tourism and Himalayan Biodiversity: Souvenir and Abstracts*. Govt. PG College, Uttarkashi.

Singh, V., Gaur, R.D. and Bohra, Babita. 2008. A Survey of Fodder Plants in Mid-altitude Himalayan Rangelands of Uttarakhand, India. *Journal of Mountain Science*, 5(3): 265-275.

Chapter 11

Nutrient Utilization of Mustard Straw Supplemented with Different Levels of Green Oat Fodder in Crossbred Heifers

☆ *D.N. Pandey, Ashoka Kumar, Ripusudan Kumar and Vir Singh*

ABSTRACT

Mustard straw is an un-conventional fibrous feed for ruminants. Evaluation of the nutrient utilization of mustard straw with different ratios of green oat fodder was done using crossbred heifers. In a randomized block design 16 crossbred heifers were assigned to evaluate the 4 different ratios of mustard straw and green oat fodder. A 30 days' feeding trial with 5 days digestion trial was carried out to determine the nutrient intake and digestibility. Significantly ($P<0.05$) increased DM and cellulose contents were observed from T1 to T4. OM, CP and lignin contents for all treatments were found similar. T1 showed significantly ($P<0.05$) lower NDF and ADF values. Hemi-cellulose content was decreased from T1 to T4. Intake of different nutrients was significantly ($P<0.05$) higher in T1 and lower in T4. ADF and cellulose intake were not significantly different among the treatments. Similar to the intake, digestibility of nutrients was also higher in T1 and lower in T4. Intake of DDM, DOM, DCP and TDN was significantly ($P<0.05$) higher in T1 and lower in T4, both in terms of g/d and g/kg$W^{0.75}$. On the basis of present results it can be concluded that mustard straw and green oat fodder at 50:50 level is better in terms of composition, intake and digestibility of nutrients for crossbred heifers.

Keywords: *Mustard straw, Oat fodder, Heifers, Nutrients intake and Digestibility.*

Introduction

Due to fast increasing human population in India, more and more land is being used for producing food grains, resulting in lesser land available for grazing and fodder production. Therefore, more efficient utilization of existing feed resources, particularly of unconventional ones is urgently warranted. Mustard (*Brassica juncea*) is the second most important group of oil seeds and India is the second largest producer of mustard in the world (Kiresur, 1999), producing 20 per cent of the world rapeseed mustard. Mustard group oil seeds crops have the advantage of being more drought tolerant and disease resistant than other oil seeds crops and, with the recent increases in renovated dry land agriculture in India, the production of mustard has been increased substantially.

Presently in India, around 14 million tones of mustard straw are being produced annually (Mishra *et al.*, 2000). In the crop of mustard, oil cake and meals are commonly used in livestock production, but the rest part of plant, *i.e.* stem and leaves (straw) is usually a waste material remaining un-utilized for any substantive purpose. It has emerged as an important alternative feed source for livestock population. Due to the poor digestibility of mustard straw, animals consume it in lesser amounts (Mishra *et al.*, 2000). The intake and nutrient utilization can be improved with so many treatments and/or supplemented with green fodders. Oat (*Avena sativa* L.) is an important non-legume *rabi* (winter season) cereal cultivated extensively for forage and grain. Its green yield may go up to 500-600 quintal per hectare (Singh, 2001). It is highly nutritive forage for livestock with 11.4 per cent CP, 0.51 per cent Ca and 0.28 per cent phosphorus (Prakash *et al.*, 1997). It ranks sixth in the world agricultural production after wheat, rice, maize, barley and cotton. Because of excellent growth potential and quick re-growth, oat has wide adaptability in Northern region of the country. Attempts have been made to improve mustard straw supplemented with various level of green oat fodder for nutrient utilization in crossbred heifers.

Materials and Methods

The present experiment was conducted on four different combination of mustard straw and green oat fodder, *i.e.* 50:50 (T1), 60:40 (T2), 70:30(T3) and 80:20 (T4), respectively.

Selection of Animals and Experimental Design

Sixteen growing crossbred heifers of body weight ranging from 43-130 kg were selected from the general herd of the University Dairy Farm and these were equally distributed in to four treatments of each with four animals following Randomized Block design (RBD).

Housing and Management of the Animals

Animals were housed in a well ventilated, dry and pucca shed having individual mangers for feeding. All the heifers were let loose in an open shed for exercise in the morning from 8AM to 10AM during the adaptation periods and animal had access to clean and fresh drinking water taking all hygienic measures. All the animals were vaccinated against common epidemic diseases and dewormed before the start of the experiment. During the period of digestion trial animals kept were confined only to the shed. Before and after the digestion trial, body wt. of animals was recorded.

Feeds and Feeding Schedule

In treatment first, animal were fed mustard straw and green oat fodder in the ratio of 50:50 on as such basis, the ratios were changed to 60:40, 70:30 and 80:20 in treatment second, third and fourth, respectively. Experimental fodders were offered in the forenoon at 10AM, after chopping them into pieces of 2.5-3 cm size. Experimental feeding was done for a period of 21 days, followed by five days digestion trial.

Digestion Trial

After the end of 21 days of experimental feeding, a digestion trial was conducted for five days with three days as adaptation period to estimate intake and utilization of nutrients. During that period, intake was estimated by weighing the amount of feed offered to and residue left by each animal and total collection of feaces was done regularly for 24 hours.

Collection, Sampling and Preservation of the Sample

The representative samples of feed ingredients, different combinations of experimental feed with maintaining the above ratios, residue left and feaces sample were collected for five days for the determination of nutrient intake and also pooled for further various chemical analyses. Suitable aliquot of faeces samples were also preserved for N-estimation in a glass bottle after mixing with 25 per cent sulphuric acid for five days.

Analytical Methods

All samples were analyzed for their DM, OM, CP, TA, GE and cell wall constituents such as NDF, ADF, hemi-cellulose, cellulose and lignin. DM, OM, CP, TA were determined following the methods of AOAC (1995). However, NDF, ADF and lignin contents were determined by the method of Goering and Van Soest (1970), whereas hemi-cellulose and cellulose contents of the feed were worked out by differences between NDF to ADF and ADF to lignin, respectively. The gross energy was determined by the chromic acid oxidation of O'Shea and Maguire (1962) method. The data were analyzed statistically as per Snedecor and Cochran (1970).

Results and Discussion

Chemical Composition of Feeds

DM, OM, CP, TA, NDF, ADF, GE, hemi-cellulose, cellulose and lignin contents for different combinations of mustard straw and green oat fodder have been presented in Table 11.1. Results showed that the DM content increased significantly ($P<0.05$) from T1 (55.44 per cent) to T4 (72.82 per cent) and the difference between these combinations was 16.38 units. This variation among the different fodder combinations might be due to increasing proportion of mustard straw in the diets. Malik (2000) observed similar trend in the combination of paddy straw plus lahi (mustard). OM, CP, GE and lignin contents of all treatment were almost similar and having no significant difference between the treatments. T1 showed significantly lower NDF and ADF values (67.30 and 46.55 per cent ,

Table 11.1: Chemical Composition of Different Combinations of Mustard Straw and Green Oat Fodder

Particulars	*T1 Mustard Straw : Green oat (50 : 50)*	*T2 Mustard Straw : Green Oat (60 : 40)*	*T3 Mustard Straw : Green Oat (70 : 30)*	*T4 Mustard Straw : Green Oat (80:20)*
DM	55.44±0.21[d]	61.50±0.40[c]	66.78±0.19[b]	72.82±0.31[a]
OM	91.94±0.38	91.72±0.27	91.66±0.21	92.04±0.48
CP	8.76±0.22	8.39±0.27	8.77±0.21	8.39±0.10
GE	4.44±0.19	4.73±0.17	4.41±0.28	4.38±0.19
NDF	67.30±0.28[b]	67.49±0.49[ab]	68.27±0.68[ab]	6[illegible].22±0.33[a]
ADF	46.55±0.43[d]	48.31±0.29[c]	51.38±0.17[b]	53.7[illegible]±0.25[a]
Hemi-cellulose	20.75±0.15[a]	19.19±0.21[a]	16.89±0.51[b]	15.47±[illegible]58[b]
Cellulose	38.27±0.52[d]	40.43±0.60[c]	43.18±0.39[b]	45.48±0.4[illegible]
Lignin	8.28±0.09	7.88±0.32	8.20±0.22	8.28±0.18

a, b, c, d: Figures bearing different superscripts in a row differ significantly ($P<0.05$).

respectively) and higher NDF and ADF values were found in T4 (69.22 and 53.75 per cent , respectively). Hemi-cellulose content decreased from T1 (20.75 per cent) to T4 (15.47 per cent) but the cellulose content increased from T1 (38.27 per cent) to T4 (45.48 per cent), indicating that the mustard straw contained less hemi-cellulose and higher cellulose contents than green oat fodder. The results of NDF and ADF content found are in present study is agreement with the values reported by Pradhan (1976). Malik (2000) reported the lowest hemi-cellulose content (15.4 per cent) in paddy straw + lahi at 20:80 as compared to 24.4 per cent at 80:20 ratios.

Nutrient Intake

The intake of different nutrients measured in kg/day or g/day, kg/100kg b. wt. and per unit of metabolic b. wt. (g/kgW$^{0.75}$) has been shown in Table 11.2. The mean intake (kg/day) of dry matter was 3.13, 2.58, 1.88 and 1.66 in T1, T2, T3 and T4, respectively. Corroborative results were observed by Sohane and Singh (2001) who reported that the intake of nutrients significantly increased with supplementation of green fodders with rice straw from two varieties compared to sole feeding of rice straw. When the intake of dry matter was calculated on the basis of 100-kg body wt. and metabolic body wt. (g/kgW$^{0.75}$) it came to be 4.13, 3.68, 2.45, 2.13 and 119.65, 108.07, 71.99 and 63.05 in T1, T2, T3 and T4, respectively. Brid *et al.* (1994) reported that the intake of wheat straw increased with the increased level of green Lucerne in the diets. That result is comparable with present findings. The total dry matter intake (kg/day) was significantly ($P<0.05$) higher in T1 and lower in T4, but when the intake was revealed as kg/100kg body wt., T1 and T2 were found significantly ($P>0.05$) similar and higher than T3 and T4, which was significantly ($P>0.05$) similar. Saccharise *et al.* (1995) reported that the voluntary intake for dry matter of Ethiopian mustard (*B. carianta*) ranged from 30.61 to 44.84 g/kgW$^{0.75}$, the value was lower than that obtained in the present study. However, Demorquilly (1992) reported that the intake of rape fodder was 120 g/kgW$^{0.75}$, which was slightly higher than the value of the present study. Similar to DMI (kg/100kg b.wt.), OMI was found significantly ($P>0.05$) similar in T1 and T2, which were higher than in T3 and T4. The higher dry matter intake in T1, resulted in increased organic matter intake (OMI), crude protein intake (CPI), neutral detergent fiber intake (NDFI), acid detergent fiber intake (ADFI), hemi-cellulose intake and

Table 11.2: Intake of Different Nutrients in Different Treatments

Attributes	*T1*	*T2*	*T3*	*T4*	*SEM±*
No. of animal	4	4	4	4	
Mean b. wt. (kg)	79.25±20.05	79.50±17.10	79.0±12.43	79.50±8.92	6.81^{NS}
Metabolic b.wt. ($kgW^{0.75}$)	26.08±5.03	26.32±4.14	26.31±3.15	26.53±2.25	1.70^{NS}
DMI kg/day	3.13^{a}±0.62	2.58^{ab}±0.48	1.88^{ab}±0.19	1.66^{b}±0.07	0.24
Kg/100kg b.wt	4.13^{a}±0.26	3.68^{a}±0.14	2.45^{b}±0.19	2.13^{b}±0.15	0.23
$g/kgW^{0.75}$	119.65^{a}±1.79	108.07^{b}±1.98	71.99^{c}±3.38	63.05^{d}±2.64	6.23
OMI kg/day	2.88^{a}±0.58	2.62^{ab}±0.44	1.72^{ab}±0.18	1.52^{b}±0.07	0.22
Kg/100kg b.wt	3.80^{a}±0.23	3.37^{a}±0.13	2.25^{b}±0.17	1.96^{b}±0.14	0.21
$g/kgW^{0.75}$	109.98^{a}±1.57	99.07^{b}±1.83	65.94^{c}±3.07	57.99^{d}±2.38	5.72
CPI g/day	274.25^{a}±54.45	240.0^{ab}±40.44	164.25^{ab}±17.15	139.0^{b}±6.47	21.15
$g/kgW^{0.75}$	10.48^{a}±0.16	9.08^{b}±0.17	6.30^{c}±0.30	5.29^{d}±0.21	0.55
NDFI (g/day)	2106.8^{a}±418.32	1928.5^{ab}±324.6	1280.3^{ab}±132.7	1145.8^{b}±51.8	161.99
ADFI (g/day)	1456.8±289.44	1380.3±232.45	963.3±99.90	889.8±40.22	107.7^{NS}
HCI (g/day)	649.5^{a}±129.15	565.3^{ab}±108.7	316.8^{bc}±32.79	256.3^{c}±11.65	57.37
Cellulose I(g/day)	1198±237.72	1155.3±194.49	809.8±83.99	752.8±34.15	88.18^{NS}

a, b, c, d Figures bearing different superscripts in a row differ significantly ($P<0.05$).

NS: Not significant.

cellulose intake in the same treatment. This might be due to inclusion of more green fodder (oat) in the diet, which is more palatable than dry sources of roughages, like mustard straw (Mosi and Butterworth, 1985).

Digestibility of Nutrients

Digestibility of different nutrients has been presented in Table 11.3. The dry matter digestibility values of T1, T2, T3 and T4 were 62.36, 55.24, 44.66 and 42.08 percent, respectively, which significantly ($P<0.05$) differed between the treatments. The reason for decrease in dry matter digestibility from T1 to T4 might be due to increased amount of mustard straw in the diet, which results in more cell wall constituents in the diet. It has been suggested by Orpin and Bountiff (1978) that forage supplements containing significant quantities of soluble nutrients may influence the rate of fiber digestion in the rumen by stimulating colonization of plant material by rumen microorganism. The *in-sacco* digestibility of untreated and ammoniated straw significantly increased by inclusion of Lucerne in the diets. Stutchmare and Martin (1972) observed that fiber contents, *i.e.* NDF, ADF and ADL were negatively correlated with dry matter digestibility. Dry matter digestibility can vary widely according to the diet of animal (Ranjhan, 1980).

The digestibility of all other nutrients, like OM, CP, NDF, ADF, HC, Cellulose and GE were significantly ($P<0.05$) higher in T1 and lower in T4 compared to other treatments. Sohane and Singh (2001) also reported that the digestibility of nutrients significantly increased with supplementation of green fodders with two rice straw varietes compared to sole feeding of rice straw. Digestibility of nutrients reduced from T1 to T4 due to the increased proportion of mustard straw to green oat fodder. The differences among the treatments may be due to the high lignin content (11.50 per cent) in mustard straw, which showed negative relation with organic matter digestibility (Nadia *et al.*, 1973). Armstrong (1993) reported the higher value for organic matter digestibility of rape forage (86.5 per cent) than found in present study. Das and Singh (1999) reported increased *in-vitro* DM and OM digestibility with supplementation of increased proportion of berseem in wheat straw based diet, which is similar to present study. Results of CP digestibility in present study have been in agreement with those reported, that the apparent digestibility of CP increased with increased supplementation of green forage. Setty

Table 11.3: Digestibility of Nutrients in Different Treatments

Attributes	*T1*	*T2*	*T3*	*T4*	*SEM±*
DM	$62.37^{a}\pm1.39$	$55.25^{b}\pm0.55$	$46.17^{c}\pm0.96$	$42.09^{d}\pm0.35$	2.08
OM	$70.19^{a}\pm0.80$	$66.36^{b}\pm1.82$	$54.62^{c}\pm0.83$	$47.95^{d}\pm0.60$	2.36
CP	$57.97^{a}\pm1.07$	$52.78^{b}\pm0.75$	$44.30^{c}\pm0.57$	$41.21^{d}\pm0.54$	1.75
GE	$66.07^{a}\pm0.59$	$62.04^{b}\pm0.47$	$59.66^{c}\pm0.61$	$54.61^{d}\pm0.65$	1.10
NDF	$58.61^{a}\pm0.79$	$47.99^{b}\pm1.18$	$43.81^{c}\pm1.36$	$40.25^{d}\pm0.54$	1.84
ADF	$47.98^{a}\pm0.93$	$43.99^{b}\pm0.80$	$41.94^{b}\pm0.38$	$37.70^{c}\pm0.47$	1.00
Hemi-cellulose	$66.96^{a}\pm0.56$	$61.05^{b}\pm0.75$	$57.42^{c}\pm0.77$	$55.33^{d}\pm0.37$	1.17
Cellulose	$53.61^{a}\pm0.33$	$49.75^{b}\pm0.48$	$47.43^{c}\pm0.40$	$46.11^{d}\pm0.46$	0.76

a, b, c, d Figures bearing different superscripts in a row differ significantly ($P<0.05$).

Table 11.4: Plane of Nutrition in Different Treatments

Attributes	*T1*	*T2*	*T3*	*T4*	*SEM±*
DDMI (g/day)	1972.8^{a}±425.1	1583.5^{ab}±276.5	866.2^{bc}±92.7	696.6^{c}±32.78	177.2
$g/kgW^{0.75}$	74.69^{a}±2.65	59.73^{b}±1.66	33.14^{c}±0.96	26.54^{d}±1.15	5.11
DOMI (g/day)	2028.4^{a}±414.91	1751.5^{ab}±323.17	936.71^{bc}91.23	729.1^{c}±28.06	184.13
$g/kgW^{0.75}$	77.21^{a}±1.73	65.84^{b}±2.96	35.98^{c}±1.51	27.83^{d}±1.36	5.35
DCPI (g/day)	158.59^{a}±30.65	127.14^{ab}±22.35	72.88^{bc}±7.89	57.35^{c}±3.17	13.66
$g/kgW^{0.75}$	6.08^{a}±0.18	4.79^{b}±0.15	2.79^{c}±0.11	2.18^{d}±0.07	0.41
$^{+}$TDNI (g/day)	2129.85^{a}±435.7	1839.06^{ab}±339.3	983.54^{bc}±95.8	765.60^{c}±29.5	193.34
$g/kgW^{0.75}$	81.07^{a}±1.82	69.13^{b}±3.11	37.78^{c}±1.59	29.22^{d}±1.42	5.61

+ TDN calculated from DOM (1kg DOM = 1.05kg TDN; NRC, 1989).

a, b, c, d Figures bearing different superscripts in a row differ significantly ($P<0.05$).

(1973) observed highly negative correlation between CP digestibility and increased level of dry fodder inclusion in the diet. The ADF digestibility was significantly ($P<0.05$) lower (37.20 per cent) in T4 and higher (47.90 per cent) in T1 due to changed proportion of mustard straw and green oat fodder. Results of present study indicated that ADF digestibility increased with increased green fodder supplementation, similar findings were observed by Reddy (1998) and Dutta *et al.* (1999).

Digestible Nutrient Intake/Plane of Nutrition

The intake of digestible dry matter (DDMI), digestible organic matter (DOMI), digestible crude protein (DCPI) and total digestible nutrients (TDNI) in terms of g/day and $g/kgW^{0.75}$ have been presented in Table 11.4. The DDMI, DOMI, DCPI and TDNI were significantly ($P<0.05$) higher in T1 and lower in T4, both in the terms of g/day and $g/kgW^{0.75}$. The increase in intake in the oat fodder supplemented diets was probably due to enhanced cellulolysis, digestion of cell wall in the reticulo-rumen (Khandeker *et al.*, 1998). The results of present study are in agreement with the results of Dutta *et al.*, (1999), who reported that the intake of DCP and DOM were significantly higher in tree forage supplemented diet compared to only rice straw based diet in goats.

Conclusion

The results of this study suggest that increasing of mustard straw in the diet can decrease the intake and digestibility of nutrients. It can be concluded that mustard straw and green oat fodder at 50:50 ratio is better in terms of composition, intake and digestibility of nutrients for crossbred heifers. Mustard straw often ends into waste in India. Its utilization as a component of animal feed would help ease out the problem of feed scarcity to a certain extent to the benefit of farmers.

Acknowledgement

This chapter is based on the thesis research work of the first author. Financial support by Indian Council of Agricultural Research, New Delhi, India is gratefully acknowledged. Facilities for experiments provided by Director Research, GB Pant University of Agriculture & Technology, Pantnagar, India are also acknowledged.

References

AOAC, 1995. *Official Method of Analysis, Association of Official Analytical Chemist International*, Vol. 1, (16th ed.). Edited by Patricia Cunniff, Virginia 2201-3301, USA.

Armstrong, R.H., M.M. Beattie and E. Robertson, 1993. Intake and digestibility of components of forage rape (*Brassica rapus*) by sheep. *Grass and Forage Sci.* 48(4):410-415.

Brid, B., S.H. Romulo, and R.A. Leng, 1994. Effect of lucern supplementation and defaunation on feed intake, digestibility, N-retention and production of sheep, fed straw based diets. *Ani. Feed Sci. and Tech.* 45(6):119-129.

Das, A. and G.P. Singh, 1999. Effect of different proportions of berseem and wheat straw on in-vitro digestibility and total gas production. *Indian J. of Anim. Nutr.* 16(1):60-64.

Demarquilly, C. and J. Andrieu, 1992. Chemical composition, digestibility and intake of green European forages. *Production Animals.* 5(3):213-221.

Dutta, N., K. Sharma, and Q.Z. Hassan, 1999. Effect of supplementation of rice straw with *Leucaena leucocephala* and *Prosopis cineraria* leaves on nutritive utilization by goats. *Asian-Aus. J. Anim. Sci.* 12(5):742-746.

Goering, H.K. and P.J. Van Soest, 1970. Forage fibre analysis. Agriculture Handbook No. 379: 1-12 Agriculture Research Services, United States, Deptt of Agriculture.

Khandeker, Z.H., H. Steingass and W. Dcochner, 1998. supplementation of wheat straw with sesbania on voluntary intake, digestibility and rumen fermentation in sheep. *Small Ruminant Res.* 28:23-29

Kiresur, V.R., 1999. The Yellow Revolution. *Employment News*, 31 July-6 August, pp,1-2.

Malik, P.K., 2000. Supplementation of rice straw with different levels of green lahi fodder in crossbred heifers. *Thesis*, M.Sc.Ag. (Animal Nutrition) G. B. Pant Uni. of Agri. and Tech., Pantnagar. 123p.

Mishra, A.S., O.H. Chaturvedi, Ananta Khali, R. Prasad, A. Santra, A.K. Misra, S. Parthasarathy and R.C. Jakhmola, 2000. Effect of sodium hydroxide and alkaline hydrogen peroxide treatment on physical and chemical characteristics and IVOMD of mustard straw. *Ani.Feed Sci. and Tech.*, 84:257-264.

Mosi, A.K. and M.H. Butterworth, 1985. The voluntary intake and digestibility of combination of cereal crop residues and legume hay for sheep. *Ani. Feed Sci. and Tech.* 12:241-251.

Nadia, P.C., A.J. Gorden and Nevdoerffer, 1973.The effect of the chemical composition of maize plant lignin on the digestibility of maize stalk in the rumen of cattle. *Brit. J. Nutr.* 29:1-12.

O Shea, J. and M.I. Maguire, 1962. Determination of calorific value of feed stuff by chromic acid oxidation. *J. Sci. food and Agri.* 13:530.534.

Orpin, C.C. and L. Bountiff, 1978. Zoospore clematises in the rumen phycomycete *Neocallimastis frontalis*. *J.Gen. Microbiol.*104:113-122.

Pradhan, K., 1976. Chemical composition (in per cent of DM) of different legume and non-legume forages harvested at preflowering (i), flowering (ii) and post flowering (iii) stages. In: Ranjhan, S.K. ed. Animal Nutrition and Feeding Practices in India, New Delhi. Vikas Publication House. Pp339-340.

Prakash, O.; D.N. Verma and S.N. Lal, 1997. Yield and quality of various varieties of oat fodder. *Indian J. of Anim. Nutr.* 14:258-261.

Ranjhan, S.K., 1980. Animal Nutrition in the Tropics. Vikas Publishing House Pvt. Ltd., Vikas house, 2104, Industrial area, Sahibabad , Gaziabad, U.P., India.

Reddy, D.V., 1998. The effect of supplementation of green forage on utilization of rice straw-poultry dropping-Rice bran-Fish meal diets in buffalo. *Buffalo J.* 14(1):31-44.

Setty, S.V.S., 1973. Studies on nutritive value of forages. *Indian J. Dairy Sci.* 26:1-8.

Singh, R.R., I.S. Agrawal and J.S. Verma, 2001. Chemical composition and production potential of different strain of fodder oat. *Indian J. of Anim. Nutr.* 18(2):165-171.

Snedecor, G.W. and W.S.G. Cochran, 1970. *Statistical method.* The towa state university press, Arness towa, U.S.A.

Stutchmane, D.P. and C.C. Martin, 1972. Genetic variation in yield and quality of oat. *forage crop Sci.* 12:831-833.

Chapter 12

Animal Feeding: Frequently Asked Questions

☆ *Ashoka Kumar*

Q.1. *What is the importance of feeding green forages to the animals?*

Ans. Green forages are easily digestible and palatable. They contain sufficient amount of protein, vitamin A, calcium and phosphorus etc. Adequate supply of green forages reduces the cost of feed on concentrate and improves the milk quality.

Q.2. *How green forages can protect them from being dry?*

Ans. Green forages can be preserved in the form of hay and silage, which contain all the nutrients, and they can be fed to the animals when needed.

Q.3. *What are the disadvantages of feeding only dry roughages?*

Ans. Only dry roughages cannot supplies the adequate amount of nutrients required for the growth and production. Feeding of only dry roughages causes constipation and reduced milk production.

Q.4. *How can we diagnose the Vitamin D deficiency in the animals?*

Ans. Vitamin D deficiency in the animals causes some specific symptoms like- weakness, weak bones, retarded growth and difficulty in breathing.

Q.6. *How much salt should be added to the daily diet of the animals?*

Ans. 50 gm salt (NaCl) can be added to the diet of healthy animals. Chlorine deficiency cannot occur by salt feeding.

Q.7. *How can you improve the digestibility of animal feed?*

Ans. The roughage can be given by chaffing and cutting to improve the digestibility. Concentrate mixture should be fed to the animal by grinding, soaking or sometimes in the form of pellets.

Q.8 *What is the importance of colostrum feeding in newly born calves?*

Ans. Calves should be fed the colostrum within two hours of birth. The importance of receiving colostrum at birth is that it is rich in protein, vitamin A, and minerals. It also supplies antibodies and immunoglobulin that a newborn calf lacks, which improves their immunity.

Q.9. *What amount of colostrum should be fed to the calves?*

Ans. Calves should recieve about 1/10 of body weight of colostrum 3-4 times a day.

Q.10. *What precautions should be kept in the mind while feeding urea treated straw?*

Ans. Urea treated straw should be fed after 20-25 days of urea treatment. It should be spread in the air just before 15-20 minutes of feeding. Over feeding should not be done. Urea treated straw is never fed to the young animals below 6 months of age.

Q.11. *How can the urea toxicity be treated?*

Ans. In case if animal had consumed the urea and not treated then it may die due to urea toxicity. To prevent the urea toxicity ½- 1 gallon (2.5 lit.) 5 per cent vinegar should be given to the animals.

Q.12. *At which stage of pregnancy, salt feeding should be stopped in pregnant cows?*

Ans. In pregnant cows salt feeding should be stopped before 3 months of calving, otherwise there may be edema in the udders.

Q.13. *What feed should be given to the animals during drought?*

Ans. The roughages, which are preserved in the form of hay and silage, can be given to the animals during drought. Besides it, other type of fodder like sunflower leaves, green algae, and hyacinth, sugarcane tops and tree leaves can be given to the animal during drought period.

Q.14. *What are the causes of bloat? How can it be treated in cows?*

Ans. Feeding of large amount of green legumes and contaminated roughage causes the bloat. During first aid treatment a wood piece should be kept in between the jaws and immediately 60 ml. of turpentine oil, 4 ml. of phenol, 4 ml of nuxvomica and 1 lit of til oil should be given.

Q.15. *How farmers can obtain maximum production from the lactating animals?*

Ans. Farmers can obtain the maximum production by keeping following points in their mind. The diet should be balanced and adequate. The milking time should be fixed. Milking should be completed within 2-3 minutes of milk let down from the udders. Exciting should be prevented during milking. High yielding cows should be milked thrice daily.

Q.16. *What is the importance of balance diet for the animals?*

Ans. Balance diet provides all the nutrients required for the maintenance and growth, production, reproduction as well as for work. If the diet is not balanced with all the nutrients then it can affect production and increase the susceptibility to diseases.

Q.17. *Is it sufficient to feed only green or dry fodders to fulfill the nutrients requirement of the animals?*

Ans. Only green fodder cannot fulfill all the requirements of nutrients of the lactating and pregnant animals, as they do not contain adequate amount of nutrients required for the growth and production. So they require adequate amount of balanced concentrate mixture in addition to green or dry fodder.

Q.18. *Formulate the ration for a 450 kg buffalo, producing 10 kg milk daily, by which she can maintain its production for a longer time.*

Ans. A lactating buffalo require 3 kg dry matter/100kg b.wt per day from roughage and concentrate mixture. According to this a 450 kg buffalo will require 13.5 kg DM. About 5 kg concentrate mixture for milk production and 1.5 kg to maintain body wt is required which makes about 6.5 kg concentrate mixture. Remaining 7 kg DM can be supplied by feeding 8 kg wheat/paddy straw or 28-30 kg green fodder.

Q.19. *Sometimes high producing buffaloes suddenly fall down and cannot standup. If not treated then die. Why?*

Ans. This disease occurs due to Ca deficiency, which is known as milk fever. Milk contains a large amount of Ca, so in high producing animals Ca deficiency takes place. Feeding balanced concentrate mixture, which contains adequate amount of Ca, and injection of Calcium borogluconate can prevent this disease.

Q.20. *What are the symptoms of nutrients deficiency in animals?*

Ans. Usually animals become weak. Other symptoms are reduced appetite and lowered feed intake, lower milk production; reproduction efficiency goes down and susceptibility to diseases is increased.

Q.21. *How does balance concentrate mixture improve the resistance power to diseases?*

Ans. Balanced concentrate mixture contains protein fat, 25 per cent mineral mixt. and 15 per cent vitamins, which can improve the health of animals. The good health improves the resistance power of animals to diseases.

Q.22. *Do the balanced concentrate mixture affect the period of estrous and conception?*

Ans. Yes, the balanced concentrate mixture contains some mineral elements, like Cu, P, Zn etc. which can affect the reproduction and helps to come in heat and conception.

Q.23. *Does the balanced concentrate mixture reduce the dry period in lactating animals?*

Ans. Yes, feeding of balanced concentrate mixture produces the milk for a longer period and animals become pregnant after 3 months of parturition. The farmers should stop the milking just before 2 months of next calving, by which they can produce more milk in next lactation.

Q.24. *At which age feeding of green forage and concentrate mixture should be started to calves?*

Ans. Only milk feeding should be done up to 2 months of age of the calves @ 1/10 per cent of body weight. Soft green forage can be offered at 15 to 20 days age.

Q.25. *Usually the farmers give the mustard oil to their calves. Is it correct or not?*

Ans. No, it is not a good practice because it may cause the diarrhoea and pneumonia and sometimes calves may die. So they should provide only colostrums.

Q.26. *What are the methods of feeding of concentrate mixture to lactating animals?*

Ans. The amount of concentrate mixture depends upon milk production capacity of the animals. This ration should be fed twice in a day. And it should be given at milking time. The feed can be given in dry form or in the form of sani.

Q.27. *Usually the animals consume large amount of green fodder in rainy season and suffer with bloat. What is the treatment at that time?*

Ans. In this case feeding of green fodder should be completely stopped. Only dry bhusa should be fed to the animal and consult with the doctor.

Q.28. *Is it necessary to feed concentrate mixture to the goats?*

Ans. Yes, The adult lactating goat should be fed about 1/3 of concentrate mixture of daily milk yield.

Q.29. *Is green forage sufficient to maintain milk-yielding capacity of lactating animals?*

Ans. It depends upon the variety of roughages. Berseem, lobia etc. are good quality fodders, which are able to maintain 5-8kg milk production by feeding *ad lib.* But feeding of only green oat can only fulfill the nutrient requirement for 3-4 kg milk production. Then high producing animals require balanced concentrate mixture to get maximum production.

Q.30. *What is the importance of mineral mixture for the animals?*

Ans. Mineral mixture is required for the proper metabolic activities of the animal body. It is required for strong and proper growth of bones. It also improves the milk production and maintains reproductive ability.

Index

D

E

F

T